RESEARCH IN MARITIME HISTORY
NO. 21

THE EXPLOITED SEAS:

New Directions for
Marine Environmental History

Edited by

Poul Holm, Tim D. Smith and David J. Starkey

International Maritime Economic History Association/
Census of Marine Life

St. John's, Newfoundland
2001

ISSN 1188-3928
ISBN 0-9730073-1-1

Research in Maritime History is available free of charge to members of the International Maritime Economic History Association. The price to others is US$15 per copy, plus $3.50 postage and handling.

Back issues of *Research in Maritime History* are available:

No. 1 (1991)   David M. Williams and Andrew P. White (comps.), *A Select Bibliography of British and Irish University Theses about Maritime History, 1792-1990*

No. 2 (1992)   Lewis R. Fischer (ed.), *From Wheel House to Counting House: Essays in Maritime Business History in Honour of Professor Peter Neville Davies*

No. 3 (1992)   Lewis R. Fischer and Walter Minchinton (eds.), *People of the Northern Seas*

No. 4 (1993)   Simon Ville (ed.), *Shipbuilding in the United Kingdom in the Nineteenth Century: A Regional Approach*

No. 5 (1993)   Peter N. Davies (ed.), *The Diary of John Holt*

No. 6 (1994)   Simon P. Ville and David M. Williams (eds.), *Management, Finance and Industrial Relations in Maritime Industries: Essays in International Maritime and Business History*

No. 7 (1994)   Lewis R. Fischer (ed.), *The Market for Seamen in the Age of Sail*

No. 8 (1995)   Gordon Read and Michael Stammers (comps.), *Guide to the Records of Merseyside Maritime Museum, Volume 1*

No. 9 (1995)   Frank Broeze (ed.), *Maritime History at the Crossroads: A Critical Review of Recent Historiography*

No. 10 (1996)   Nancy Redmayne Ross (ed.), *The Diary of a Maritimer, 1816-1901: The Life and Times of Joseph Salter*

No. 11 (1997)   Faye Margaret Kert, *Prize and Prejudice: Privateering and Naval Prize in Atlantic Canada in the War of 1812*

No. 12 (1997)   Malcolm Tull, *A Community Enterprise: The History of the Port of Fremantle, 1897 to 1997*

No. 13 (1997)   Paul C. van Royen, Jaap R. Bruijn and Jan Lucassen, *"Those Emblems of Hell"? European Sailors and the Maritime Labour Market, 1570-1870*

No. 14 (1998)   David J. Starkey and Gelina Harlaftis (eds.), *Global Markets: The Internationalization of The Sea Transport Industries Since 1850*

No. 15 (1998)   Olaf Uwe Janzen (ed.), *Merchant Organization and Maritime Trade in the North Atlantic, 1660-1815*

No. 16 (1999)   Lewis R. Fischer and Adrian Jarvis (eds.), *Harbours and Havens: Essays in Port History in Honour of Gordon Jackson*

No. 17 (1999)   Dawn Littler, *Guide to the Records of Merseyside Maritime Museum, Volume 2*

No. 18 (2000)   Lars U. Scholl (comp.), *Merchants and Mariners: Selected Maritime Writings of David M. Williams*

No. 19 (2000)   Peter N. Davies, *The Trade Makers: Elder Dempster in West Africa, 1852-1972, 1973-1989*

No. 20 (2001)   Anthony B. Dickinson and Chesley W. Sanger, *Norwegian Whaling in Newfoundland: The Aquaforte Station and the Ellefsen Family, 1902-1908*

Research in Maritime History would like to thank Memorial University of Newfoundland for its generous financial assistance in support of this volume.

CONTENTS

CONTRIBUTIONS

ABOUT THE EDITORS

POUL HOLM <pho@hist.sdu.dk> is Professor of Maritime and Regional History at the University of Southern Denmark. He graduated from the University of Aalborg and earned his doctorate at the University of Aarhus. He is currently chairman of the Danish Research Council for the Humanities. He served as President for the Association for the History of the Northern Seas (1993-1997) and is currently a co-editor of the *International Journal of Maritime History* and a member of the Scientific Steering Committee for the Census of Marine Life. His research interests are mainly in the fields of marine environmental and maritime social and economic history from medieval to modern times.

TIM D. SMITH <tsmith@whsun1.wh.whoi.edu> is a fishery biologist and sometime historian of his profession, having worked for the US National Marine Fisheries Service for most of his career. His areas of specialty include population assessment and modelling, with a special concern for the effects of the exploitation of marine mammals and fish. His most recent responsibilities have included leading the US role in the work of the Scientific Committee of the International Whaling Commission. He is the author of *Scaling Fisheries: The Science of Measuring the Effects of Fishing, 1866-1955* (Cambridge, 1994). More recently he has begun a number of studies on aspects of the history of the fisheries themselves as part of the development of the History of Marine Animal Populations project.

DAVID J. STARKEY <D.J.Starkey@hull.ac.uk> is Wilson Family Senior Lecturer in Maritime History, and Director of the Maritime Historical Studies Centre, at the University of Hull, UK. He has published extensively on British trade, shipping, shipbuilding, fisheries and privateering, and is co-editor of the *International Journal of Maritime History*. His is co-editor of *England's Sea Fisheries: The Commercial Sea Fisheries of England and Wales since 1300* (London, 2000); *Global Markets: The Internationalization of the Sea Transport Industries since 1850* (St. John's, 1998); *Exploiting the Sea: Aspects of Britain's Maritime Economy since 1870* (Exeter, 1998); and *The New Maritime History of Devon* (2 vols., London, 1993); and author of *British Privateering Enterprise in the Eighteenth Century* (Exeter, 1990).

CONTRIBUTORS

YAROSLAVA ALEKSEEVA < lala_al@hotbox.ru > is a researcher at the P.P. Shirshov Institute of Oceanology at the Russian Academy of Sciences in Moscow. A specialist in marine environmental history and fish biology, she is the author of "Influence of Solovetsky Monastery Economic Activity on Ichthyofauna in the Lakes of Solovetsky Archipelago, White Sea," in *Trudy Rossiskogo Nauchno- issledovatel'skogo instituta kulturnogo i prirodnogo naslediia* (Moscow, 2002, forthcoming); and "The White Sea Fisheries from the End of the 18th Century to the 1920s," *Proceedings of the Conference "150 Years of Scientific-Fishery Research in Russia"* (Moscow, 2002, forthcoming).

DANIEL ALEXANDROV < d_alexandrov@eu.spb.ru > is Associate Professor. in the Departments of History, Political Science and Sociology at the European University in St. Petersburg, where he works mainly on the history and sociology of science and education. His recent publications include "Fritz Ringer, German Mandarins and Russian Academics," *Novoe Literaturnoe Obozrenie* LIII (2002), 24-33; and "Science in the Time of Crisis: Russia, Germany, and the USA Between Two World Wars," in V. Orel (ed.), *Nauka i Bezopasnost* (Moscow, 2000), 288-325.

MAIBRITT BAGER < Bager@hist.sdu.dk > is a history graduate of the University of Aarhus and is currently completing her doctoral studies at the University of Southern Denmark. She was formerly on the staff of the Fisheries and Maritime Museum in Esbjerg and served for several years as Assistant Book Review Editor of the *International Journal of Maritime History*. She has published previously on early modern Danish fisheries.

SEAN T. CADIGAN < scadigan@mun.ca > teaches in the Department of History and is Director of the Public Policy Research Centre of Memorial University of Newfoundland. The author of *Hope and Deception in Conception Bay: Merchant-Settler Relations in Newfound-land, 1785-1855* (Toronto, 1995), among many other works, he is currently engaged in research on forestry development and ecosystem health in coastal communities, as well as community-based fisheries management.

RUSLAN DAVYDOV < felix@dvina.ru >, a researcher at the Institute for the Ecological Problems of the North at the Ural Branch of the Russian Academy of Sciences in Arkhangelsk, is a specialist in the regional history of the Russian North. His recent publications include "The Murman Fisheries: Problems of Reliability of Data on Catches in the 19th and the Beginning of the 20th Century," in *Massovye istochniki otechestvennoi istorii* (Arkhangelsk, 2002, forthcoming); *Murman. Ocherki kolonizatsii* (Ekaterinburg, 1999, with G.P. Popov); and "The Guarding of Sea Resources in the North of Russia in the 19 and the Beginning of the 20th Centuries," in *Russki Sever v Dokumentakh Arkhiva* (Arkhangelsk, 1998), 51-71.

ZOYA DMITRIEVA < history@eu.spb.ru > is Associate Professor in the Department of History at the European University in St. Petersburg, where she studies the economic history of ancient Russia. Her main publications include *Nalogi v Rossii do XIX veka* (2nd ed., St. Petersburg, 2001, with S.A. Kozlov); and "Kataster (Piscovye knigi) des Russischen Reiches vom ende des 15 bis ins 17 Jahrhundert: Gegenwartige Forschungsprobleme und Forschungsmethoden," in *Archiv fur Diplomatik*, XL (1994), 171-203.

JOHN FIELD <jfield@u.washington.edu> graduated in 2001 from the University of Washington in the School of Aquatic and Fishery Sciences. He is working on his PhD at the same school, where his doctoral dissertation will examine the feasibility of creating a fisheries ecosystem plan for the Northern California Current ecosystem.

ROBERT C. FRANCIS < rfrancis@fish.washington.edu > is Professor in the School of Aquatic and Fishery Sciences at the University of Washington in Seattle. He has worked as a fishery biologist in international, US and New Zealand fisheries agencies. His research interests in recent years have focused on the interaction between fisheries and oceanographic changes in the North Pacific. He is a member of the steering group of the History of Marine Animal Populations project.

MICHAEL HAINES < M.S.Haines@hull.ac.uk > is HMAP Research Fellow at the Maritime Historical Studies Centre at the University of Hull He has published a number of articles on nineteenth-century fisheries history and contributed three chapters to David J. Starkey, *et al.* (eds), *England's Sea Fisheries: The Commercial Sea Fisheries of England and Wales since 1300* (London, 2000).

DIEGO HOLMGREN < dholmg@netzero.net > earned his PhD in 2001 from the University of Washington in the School of Aquatic and Fishery Sciences. He is currently working as a postdoctoral fellow at the University of California, Davis.

JEFFREY HUTCHINGS < jhutch@mscs.dal.ca > received his doctorate in Evolutionary Ecology from Memorial University of Newfoundland and conducted postdoctoral research at the University of Edinburgh and in the Canadian Department of Fisheries and Oceans in St. John's, Newfoundland. His research focusses on life history evolution, ecology, behaviour and conservation biology of marine and anadromous fishes. A member of the Department of Biology at Dalhousie University, he was recently named Canada Research Chair in Marine Conservation and Biodiversity.

ALEXEI KRAIKOVSKI < karl@eu.spb.ru > is a researcher in the Department of History at the European University in St. Petersburg. His specialty is the economic history of seventeenth-century Russia, with an emphasis on the trades in salt and fish. His recent publications include "The Salt Prices in Russia in the 1630s-1650s," *Journal of Salt History*, X (2002, forthcoming); "On the Problem of Comparability of the Data of the Customs Books and the Cloistral Accounting Books of the 17th Century. The case of the Vologda Customs Book and the Accounting Book of the Vologda Service of Spaso-Prilutsky Monastery for the Year 1634/35," in *Torgovlia, kupechestvo i tamozhennoe delo v Rossii 16 -18 vekov* (St. Petersburg, 2001), 24-30; and "The Salt Market of Totma in the 1630s and 1640s," in *Problemy social'nogo i gumanitarnogo znaniia* (St. Petersburg, 2000), 69-90.

DMITRY LAJUS < dlajus@popbio.zin.ras.spb.ru > is Senior Researcher at the White Sea Biological Station of the Zoological Institute of the Russian Academy of Sciences in St. Petersburg. His work is mainly in the field of fish population biology and ecology. His recent publications include "Long-term Discussion on the Stocks of the White Sea Herring: Historical Perspective and Present State," *ICES Marine Sciences Symposia Series*, CCXV (2002, forthcoming); "Variation Patterns of Bilateral Characters: Variation among Characters and among Populations in the White Sea Herring (*Clupea pallasi marisalbi*)," *Biological Journal of the Linnean Society*, LXXIV (2001), 237-253; and "Russia," The Status of Wild Atlantic Salmon: River-by-River Assessment," *WWF Report* (2001), 135-141 (with S. Titov and V. Spiridonov).

JULIA LAJUS <jlajus@popbio.zin.ras.spb.ru> is a researcher at the Institute for the History of Science and Technology at the Russian Academy of Sciences in St. Petersburg. She works on the history of marine biology, marine environmental history and the history of fisheries. Her recent publications include "'Foreign science' in Russian Context: Murman Scientific-Fishery Expedition and Russian Participation in Early ICES activity," *ICES Marine Sciences Symposia Series* (2002, forthcoming); "Zwischen Wissenschaft und Fisherei: Meeresforschungen im Russischen Norden am Ende des 19. und im ersten Viertel des 20. Jahrhunderts," in *Berichte zur Geschichte der Hydro- und Meeresbiologie und weitere Beitrage der 8. Jahrestagung der Deutschen Gessellssschaft fur Geschichte und Theorie der Biologie* (Berlin, 2000), 61-73; and "Science, Politics and Practice in the Fishery: Scientists, Industrialists and Fishermen in the Russian North, 1898-1940," in Poul Holm and David J. Starkey (eds). *Technological Change in the North Atlantic Fisheries* (Esbjerg, 1999), 49-60.

VLADIMIR LAPIN <history@eu.spb.ru> is Professor and Dean of the Department of History at the European University in St. Petersburg. His current research is on nineteenth-century Russian history and archival studies. He is the co-author of *Sovet ministrov Rossiskoi imperii v gody Pervoi mirovoi voiny. Bumagi A.N. Yakhontova* (St. Petersburg, 1999); and the editor of *Fondy Rossiskogo gosudarstvennogo istoricheskogo arkhiva. Kratkii spravochnik* (St. Petersburg, 1994).

VADIM MOKIEVSKY <vadim@ecosys.sio.rssi.ru> is a senior scientist at the P.P. Shirshov Institute of Oceanology of the Russian Academy of Sciences in Moscow. A specialist in marine ecology and natural conservation, he is also interested in human-nature interactions in coastal zones. His publications include: "Marine Protected Areas – International Experience in Creation and Management," *Zapovedniki i Nationalnye parki*, XXXI (2000), 49-51; "Importance of Marine Biological Resources for Russia," in N.N. Marfenin (ed.), *Rossia v okruzhaiuschem mire. Analiticheskaia seriia* (Moscow, 1999), 119-134 (with V.A. Spiridonov); and "Biological Resources of Marine and Freshwater Basins," in *Russkaia Arktika na poroge katastrofy* (Moscow, 1996), 93-101.

RANSOM A. MYERS <myers@mscs.dal.ca> holds the Killam Chair of Ocean Studies at Dalhousie University in Halifax, Nova Scotia. Dr. Myers' current, major research is on the meta-analysis of data from many populations. By treating each population as a realization of a natural experiment, it is possible to discover patterns in nature that have not been seen before because they are lost in the noise in the dynamics of individual populations.

TOM POLACHECK < tom.polacheck@marine.csiro.au > is a fishery scientist working at CSIRO Marine Research in Australia. His primary research interests are in population dynamics, fishery stock assessments and the development of robust fisheries management strategies. He has worked on a number of fishery resources worldwide, and his efforts for the last ten years have been focussed on the population dynamics of southern bluefin tuna.

CHRIS REID < Chris.Reid@port.ac.uk > is senior lecturer at the University of Portsmouth (UK), where he is also an associate of the Centre for the Economics and Management of Aquatic Resources (Cemare). His main research interest is the evolution of the British fishing industry, and he is co-editor of *England's Sea Fisheries: The Commercial Sea Fisheries of England and Wales since 1300* (London, 2000).

ARE STROM < astrom@u.washington.edu > graduated in 2001 from the University of Washington in the School of Aquatic and Fishery Sciences. He now works for Washington Department of Fish and Wildlife at the State Shellfish Laboratory in Brinnon, WA.

MALCOLM TULL < m.tull@mudoch.edu.au > is an Associate Professor in the Murdoch Business School at Murdoch University in Perth, Australia. His main research interests are in Asian economic development, maritime economics and maritime economic history. He is currently undertaking research on port privatisation and the fishing industry. He is one of the editors of the *International Journal of Maritime History*.

LANCE VAN SITTERT < cdude@beattie.uct.ac.za > is a senior lecturer in the Department of Historical Studies at the University of Cape Town. His research interest is in the field of environmental history with a special emphasis on marine environments.

ALEXEI YURCHENKO < yurchenko_alex@hotmail.com > is a Master's student in the Department of Ethnography and Anthropology at St. Petersburg State University, where he is specializing in the ethnography of the northern part of Russia, especially in fisheries culture and technology. His recent publications include "Colonization of the Barents Sea Coast from the Middle of 19th to the Beginning of the 20th Centuries: Approaches to the Economic Development of the Area," *Acta Borealia* (2002, in press); and "The Pomor's Inshore Salmon Gear at the End of the 19th and the Beginning of the 20th Centuries," *Nauka i business na Murmane. Economika i rynok* (Murmansk, 2000), 31-36.

Introduction

Poul Holm, David J. Starkey and Tim D. Smith

The current condition of many of the marine animal populations affected both directly and indirectly by the harvesting activities of human societies has been established with some certitude by aquatic and marine ecologists. Much less attention, however, has been afforded to the state of harvested populations beyond comparatively recent times, chiefly because the interests of fisheries ecologists have tended to be contemporary in scope, with less regard paid to historical trends and precedents. Their investigations, moreover, have been dominated by equilibrium-based theories of the impacts of harvesting. The reference points have too frequently been the hypothesized equilibrium state, and too seldom the long-term dynamics of population changes. At the same time, scientific studies of fisheries have often been geared to the fisheries management needs of the day, and these often have narrow regional perspectives. In this context, fisheries historians have so far made little impression. Theirs is an underdeveloped field of enquiry. As a consequence, in comparison with other fields of human production, notably agriculture and manufacturing, there are few substantial overviews of fishing, and those that have been produced adopt a national, regional or port perspective that does not elucidate the problems pertaining to marine basins and oceans. A further deficiency lies in the reluctance of fisheries historians to consider the natural environment on which fishing effort is both dependent and has a considerable impact. Historians have therefore tended to neglect the development of marine animal populations (the history of nature itself) and often lack the tools and theories to account for the fluctuations in harvests that have occurred in commercial fisheries. Historical reference points are thus important in two respects: while they offer ecologists information about the factors controlling marine populations, they also provide evidence for those historians concerned with the significance that harvesting the oceans has held for many human societies.

This volume is conceived as an attempt to correct some of these deficiencies. In combining the approaches of maritime history and ecological science it contributes to the field of environmental history, which has emerged in recent years as a distinctive branch of historical enquiry. The key to this new area of study is the integrated analysis of ecosystems and human societies, in particular the place of mankind in the historical development of ecosystems. Whereas conventional interpretations of the past adopt a markedly anthropocentric approach in explaining the developmental process, environmental historians identify mankind as but one factor in a broad ecological network of complex interactions.

Poul Holm, David J. Starkey and Tim D. Smith

The research effort in this embryonic sub-discipline has so far focussed on terrestrial environments, notably the temperate plains and tropical rain forests. Marine environmental history will broaden the scope of the research field to investigate the life forms and ecosystems that have evolved in the oceans.

Historical and Paleo-ecological Data Sources

While an abundance of historical data pertains to the extent and character of historic fish stocks and, more especially, fishing effort, the quality of the information available varies considerably according to its vintage, consistency and original purpose. These data are therefore rarely as exact as the validated data sets compiled by scientists from the records of fishing operations since the early twentieth century. In contrast, historians normally rely on information gleaned from documentary sources generated for reasons, most notably taxation, that relate to the human predator rather than the natural prey. Such data require validating in a proper historical context before they can be used to address historical and ecological questions. For instance, while taxes based on the oar or the boat generated records that indirectly reflect catches over the long term, they are generally as poor indicators of year-to-year harvests as they are *ex post facto* measures of possibilities in the sea. Other data, such as tithes or records of shares, are better indicators of yearly fluctuations but need to be corroborated. Such historical records must be carefully evaluated for use as proxy data in ecological analyses.

Potentially useful historical data have markedly different properties according to the period and purpose of their origination. Three broad chronological categories can be identified from work published by European and North American historians. First, there are national fishery-specific records that have been collected routinely, in many cases since roughly 1900, the so-called "statistical" period. These vary in terms of their character, resolution and completeness, but have been used routinely by scientists for assessing the condition of fish and other marine animal populations. These data are generally readily available in printed form and are being used increasingly by scientists to examine multi-decadal changes from the beginning of the twentieth century. Such data, however, are generally in need of validation, as data-gathering methods have differed from country to country and from decade to decade according to changes in administrative procedures. Second, an abundance of appropriate archival material, chiefly in the port and customs archives of European, American and colonial states, survives from about 1850 to 1900, what might be termed the "proto-statistical" period. This can be compared to, and synthesised with, the published data of the statistical period to extend measures of fish stocks back to the mid-nineteenth century. Third, for the period prior to about 1850, the "historical" era, data are less extensive and more difficult to interpret. Yet some historians have assembled information from internally-validated primary source materials to devise indices of catch and fishing effort that can be aligned with records from the proto-statistical and statistical periods to yield evidence suitable for use in analysing long-term changes in marine ecosystems.

In contrast to the historical records produced by humans, paleo-ecological data comes from naturally occurring ecological "archives." The character and reliability of this evidence must be determined just as carefully as that derived from documentary sources. Methods for determining ages, and for comparing consistency between parallel samples, have been developed. Further case specific development is needed, however, especially to determine temporal accuracy. Some reliable time series, notably of clupeid and anadromous species, have been developed from 1500 to the present, though the corroboration of such data with more contemporary sources is an important pending task.

As yet, neither historical nor paleo-ecological information exists for many exploited animal populations. Even where such data do exist, or can be extracted from historical and ecological archives, there are limitations inherent in, and unique to, each situation. In particular situations, however, information of potential use in improving our understanding of the history of exploited animal populations is known to exist. In such situations, further extraction and analysis will allow the development of estimates or indices of catches and fishing effort in major commercial fisheries over the period 1500-2000; the definition of more refined measures of catches and fishing effort in a few highly-regulated, high-priced inshore finfish and shellfish fisheries from 1750 onwards, based on, for example, annual reports of fishery leases; and the extension of knowledge of the proto-statistical period (1850-1900), allowing the known catch statistics to be projected back in time and thereby facilitating validation of historical period indices. Environmental history focusses on people and ecosystem structure in tandem. Applying this approach to marine animal populations, the historical, paleo-ecological and environmental data described above can be used to address questions relating to three central aspects of the subject described below.

Testing Ecological Hypotheses

The application of the broad disciplinary approach of environmental history to marine animal populations will allow robust testing of hypotheses concerning the structure and function of exploited ecosystems. In addition to the input of historians and ecologists, the involvement of physical environmental sciences is essential. The interfacing of such disciplines will allow different forms of data to be integrated, facilitating complex analyses that could not be undertaken using the techniques of the individual disciplines in isolation. This greater disciplinary breadth will improve our ability to detect the effects of abrupt environmental shocks; determine the factors contributing to gradual and cyclic changes; and explore the limitations of steady-state ecosystem models.

First, the simplest hypotheses are of abrupt environmental changes having significance to either exploited species or to species ecologically connected with them. For example, temporal changes in abundance or distribution of a species that might be reflected in historical data can easily be examined in long time series, although such retrospective correlations in themselves are not definitive. If these changes are indeed abrupt, however, as has been hypothesised for the effect of saline water intrusions on fish in the Kattegat and western Baltic, they should be relatively easy to identify in the data. Attempts to locate such factors have proven difficult, largely due to the relatively short time series of environmental and ecological data that have been available.

Somewhat more difficult to test are the second set of hypotheses, those pertaining to ecologically significant gradual or cyclic environmental changes. The classic example for exploited fish is the hypothesised effects of the long-term cycles of sunspot activity on the survival of young fish. That statistically significant correlations are easily detected in short time series is apparent in the large number of such correlations that subsequently fail as more data are collected. One limitation inherent in such tests is that the environmental variables available may themselves relate only indirectly to the real causes of ecological change. Such data may serve useful roles as proxies, but their indirect character makes inferences more difficult and less reliable.

Third, while the harvesting of individual species is usually considered in isolation, it is highly probable that interactions between species will result in more subtle long-term ecosystem change. For example, several sources of information suggest that large-scale harvesting has resulted in reductions in the size of populations at the higher trophic levels. These sources include qualitative, anecdotal observations of fish size and abundance in the popular literature from as long as 500 years ago and historical accounts of particularly large catches. Such varied pieces of evidence concur with more contemporary observations derived from developing fisheries, while modern stock assessments have demonstrated that biomass reductions have occurred in many fisheries. The scale of the reductions has generally been in the order of five to twenty times, often much more than expected, at least according to management objectives under single-species models.

Existing steady-state ecosystem models are based on assumptions of either conservation of biomass fluxes or balanced energy flows. Reductions in biomass at higher trophic levels beyond that suggested by single-species effects imply that either primary production was historically higher or that energy flows across trophic levels have changed. This indicates that historical and paleo-ecological evidence, in concert with more contemporary data, might allow us to distinguish between higher primary production or changes in energy flows.

Historical changes in primary production would suggest that alterations in environmental conditions may be at least partly responsible for the large reductions in biomass that have often been ascribed to fishing. Conversely, changes in energy flows across the trophic structure might be explained by shifts in the age and size composition of harvested populations (which are known to occur) or by changes in species composition. In these cases, historical data that provide measures of the relative contribution of fishing and environmental change to the observed biomass reductions would be useful. Changes that might be illuminated by analyses of historical and paleo-ecological evidence include shifts in relative abundance of pelagic species, alterations in relative abundance of pelagic and demersal species, and changes in relative abundance of shorter-lived species lower in the food web and longer-lived species higher in the food web. Trophic changes precipitated by the depletion of top predators may be testable by examination of historical and paleo-ecological fishery data.

Identifying *a priori* ecosystems where sufficient historical or paleo-ecological data could be extracted for testing hypotheses in these three categories will be difficult. Still, we anticipate that analyses of selected records can provide data that will enable us to resolve questions such as these in specific cases. In contrast, we do not expect to find many cases where historical or paleo-ecological records are sufficiently complete to allow the reconstruction of entire ecosystems in previous times.

The chapters that comprise this volume were first presented in February 2000 during a workshop held in Esbjerg, Denmark. This meeting was funded by the Alfred P. Sloan Foundation of New York and designed to develop a research agenda for furthering knowledge and understanding of the history of marine animal populations. The chapters are arranged in four groups, each of which illustrates particular aspects of the developing field of marine environmental history.

The first three chapters are concerned with the Newfoundland and Grand Banks' fisheries. Sources for the Newfoundland inshore, northern Labrador and offshore bank fisheries are primarily preserved in British archives, where continuing time series from the seventeenth century provide an invaluable opportunity to study the development of this important fishery. In the opening chapter, two historians, David Starkey and Michael Haines, consider primary data sources. They contend that extant statistical series may be considerably extended by the use of hitherto neglected archival records relating to the mid-seventeenth century and indicate the possibility of extending the series back even further by a sustained archival and paleo-archeological effort. A biologist, Ransom A. Myers, uses the time series to test Scott Gordon's bioeconomic model, which questions why natural renewable resource exploitation tends to become unprofitable over time. In his study of the influence of catch rates on the settlement of fishermen in Newfoundland from 1710 to 1833, Myers assumes that the number of fishermen should be an increasing function of the fishery's success rate, measured as catch per unit effort. When catch rates were good, fishermen settled. Other factors, such as wars and prices, may have affected settlement, but these are difficult to detect in the data. The fishery appears to have been close to bioeconomic equilibrium over the entire period; that is, it seems to have been fully exploited – by some definitions, over-exploited, given the type of gear and fishery practices of the time – from 1710 to 1833. The third paper of this opening section is by a biologist, Jeffrey Hutchings, and an historian, Sean Cadigan. It illustrates how marine environmental history can offer a unique perspective by interfacing the methodologies of the historical and ecological disciplines. The eighteenth and nineteenth centuries were marked by a spatial expansion of the shore, bank and Labrador fisheries. Hutchings and Cadigan demonstrate that the seemingly successful expansion of the fishery was in fact the result of the serial depletion of inshore Newfoundland cod stocks, and that the Labrador expansion had serious, negative long-term consequences for the stock of northern cod.

The archives pertaining to the Newfoundland fisheries are unique in their comprehensiveness. It is one of the prime goals of HMAP, however, to identify and generate large databases for the study of other fisheries. Each of the next three papers demonstrates the availability of historical sources and their potential for elucidating the historical and environmental questions that need to be addressed to build a history of marine animal populations. Julia Lajus, a Russian biologist, heads a team of specialists which has identified a wealth of data in the archives of Russian monasteries and state departments. The research team demonstrates the potential for reconstructing White Sea salmon populations and fishing effort from the seventeenth century onwards. For the fisheries conducted in the inshore waters of the North Sea and the Baltic, Poul Holm and Maibritt Bager likewise identify a wealth of documentary source material in Danish county records dating from the early sixteenth century. They demonstrate how the richness of the information contained in these sources may be used to address questions of environmental and historical importance, such as the periodicity of the Baltic cod and the Kattegat herring. Finally, Robert C. Francis introduces the exciting prospect of reconstructing past fish abundance by the study of fish scales in the sediment cores of the seabed. By counting and dating scales, paleo-ecologists have been able to estimate the abundance and composition of pelagic species of the Northern Californian Current over 1600 years. Francis discusses the methodological issues arising from attempts to reconstruct this ecosystem, and considers how this technique might be applied to other ecosystems.

Introduction

The third section of the volume is devoted to fisheries that have developed in the southern hemisphere in comparatively recent times, chiefly during the twentieth century. Important fisheries have been established in the upwelling areas of the Eastern Boundary Currents of the Pacific and the Atlantic. Historian Chris Reid discusses the prospects facing scholars intent on studying Central and Latin American fisheries, and naturally focuses on the anchovy fishery of Chile and Peru, which developed as the largest fishery in the world in the second half of the twentieth century. Likewise, historian Lance van Sittert considers the importance of the Benguela Current in the development of South Africa's fisheries. Both papers identify the need for much more detailed study of the information available. The Australian and New Zealand fisheries have similarly expanded rapidly through the twentieth century, and economist Malcolm Tull and biologist Tom Polacheck discuss the potential of existing historical records for studying harvesting in these areas.

The fourth section of the volume consists of Tim D. Smith's chapter, which describes a number of analyses that have sought to integrate historical and contemporary data relating to various whale populations. He focuses on the limitations of the frequently cited steady-state, or equilibrium single-species population models, used to interpret such data series and suggests that a multi-disciplinary treatment of evidence relating to these fisheries could help resolve some of the uncertainties prevalent in such examinations. In this way, the paper brings out the value that can be added to knowledge and understanding by dint of the collaboration between historians and biologists.

This message is underlined in the epilogue, which outlines how the History of Marine Animal Populations project, which has emerged from the Esbjerg workshop, aims to bring history and ecology together to build the sub-discipline of marine environmental history.

The Newfoundland Fisheries, c. 1500-1900: A British Perspective

David J. Starkey and Michael Haines

Abstract

This paper views the prosecution of the Newfoundland fishery, c. 1500-1900, from a British perspective. It considers the range, quality and utility of the evidence on the fishery in English archives and primary printed sources. The qualitative information provided by practitioners, observers and pamphleteers is examined initially. The discussion then focusses on the quantitative material relating to the fishery from the late seventeenth century onwards available in the Colonial Office papers in the Public Record Office. Some tentative results from preliminary analyses of these statistical data are presented. These indicate that the quantity of fish harvested in the eighteenth and nineteenth centuries fluctuated widely but generally increased through the period, attaining levels higher than previous estimates have shown. The catch rate is also examined to test the relationship between catch, effort and stocks. Notwithstanding the quality and volume of the primary sources, the paper concludes by suggesting that a greater research effort in the English archives will yield the data necessary to construct a rounded study of the impact of human predation on the fish stocks of Newfoundland.

Historians of the fisheries conducted from the British Isles have two main obstacles. First, this is a vast subject, due not just to its ubiquitous nature but also to the complex range of activities involved. At any given time many types of fisheries were prosecuted. While there were marked regional variations in technique, catch and market, numerous contrasting forms might be conducted concurrently from a single port or locality. Second, there are practical problems regarding the collection and interpretation of data, for the primary sources relating to Britain's fisheries are extensive, take many different forms and are scattered in repositories throughout the UK and beyond. The evidence is also uneven, both temporally (comparatively little relates to the pre-1700 period) and topically – some fisheries, and some ports, generated more, or more useful, records.[1]

---

[1]Robb Robinson and David J. Starkey, "The Sea Fisheries of the British Isles, 1376-1976: A Preliminary Survey," in Poul Holm, David J. Starkey and Jon Th. Thór (eds.), *The North Atlantic Fisheries, 1100-1976: National Perspectives on a Common*

1

The fishery and fish trade conducted by Britons (largely Englishmen) from Newfoundland illustrates these problems. While this distinctive trade has received more attention from historians than most British fishing interests, the factors governing its scale, character and significance are still imperfectly understood. This is partly due to the complexities of the political, economic, social and environmental issues that arise in examining it. But it also reflects the abundance, deficiencies and idiosyncrasies of the primary records pertaining to the business. This article surveys these problematical source materials and considers their utility in gauging the impact of human harvesting on fish populations in the waters off Newfoundland. It concludes with some tentative findings.

Fish Sources

Fishing has attracted less scholarly attention than any other of Britain's major maritime interests, perhaps reflecting its place on the national political agenda. Despite its significance in international relations, fishing has never ranked high in the list of priorities of government. In the early modern era, for instance, the state attempted to encourage the industry, but more to train seafarers for its navy than to fill the purses of fishermen or the bellies of the poor.[2] More recently, it has been claimed that British politicians have been more concerned to adhere to European directives than to cater to the interests of their own fishermen. Such neglect has been mirrored in the absence of fishing in the country's political histories. In economic terms, too, the fisheries have been overshadowed. Merely a small segment of a large, multi-faceted economy, fishing has not been as important as in states such as Iceland and Norway. This probably explains why economic historians and geographers have studied the Scottish fisheries with more enthusiasm than the English.[3] The low social status of the fisherman is perhaps a further contributory factor. Generally poor, living on the physical margins of the country, and politically dormant, these are not the labour aristocrats or class warriors who generally attract the attention of social and labour historians.[4]

Resource (Esbjerg, 1996), 121-143.

[2]Gordon Jackson, "State Concern for the Fisheries, 1485-1815," in David J. Starkey, Chris Reid and Neil Ashcroft (eds.), *England's Sea Fisheries: The Commercial Sea Fisheries of England and Wales since 1300* (London, 2000), 46-53.

[3]See, for instance, James R. Coull, *The Sea Fisheries of Scotland: An Historical Geography* (Edinburgh, 1997); and Malcolm Gray, *The Fishing Industries of Scotland, 1790-1914: A Study in Regional Adaptation* (Aberdeen, 1979).

[4]John K. Walton, "Fishing Communities, 1850-1950," in Starkey, Reid and Ashcroft (eds.), *England's Sea Fisheries,* 127-137.

If the fisheries have largely escaped the notice of British historians, this is not because of a dearth of source material. Indeed, it is more likely that the sheer abundance of evidence has deterred them, for in English archives there exists a wealth of information pertaining to the harvesting and marketing of fish in the "pre-statistical" and "proto-statistical" eras. As recent and ongoing research has indicated, a good deal of virgin information has lain undisturbed in the storage rooms of English record offices. For example, the shipping lists prepared by the Crown in the mid-fourteenth century include details of the size, ownership and home port of fishing vessels large enough to supplement Royal naval forces in time of emergency, or small enough to be exempted from such service.[5] Tithe, manorial and local customs records from the same era have been used to determine the scale and distribution of fishing operations, as well as the significance of the English fish trade.[6] Other new material that has recently come to light is qualitative in character. Evan Jones, for instance, has uncovered evidence in the English state papers that effectively revises the conventional chronology of the rise and fall of England's Iceland trade. It is commonly held that English interests in this northern trade peaked before 1497 and then declined a century or so later, due to policy changes implemented by the Danish government. Jones argues, however, that the trade was at its height in the 1490-1520 period and again in the late sixteenth and early seventeenth centuries, and contracted only in the late seventeenth century due to the imposition of salt dues.[7]

Such fresh insights bear upon the Newfoundland fishery in that they extend our knowledge of the context in which England's distant-water fishing developed. This is significant, for the ability and desire of Englishmen to exploit fish stocks in the northwest Atlantic was inextricably linked to their ability and desire to harvest fish in inshore, offshore and Icelandic waters.[8] Nevertheless, the documentary evidence concerning the Newfoundland fisheries is largely discrete from that relating to other English fisheries. It is also more voluminous, especially

---

[5]See, for instance, Great Britain, Public Record Office (PRO), C 47/2/25/18, and British Library, Add. Ms. 37494, f. 9v. We are grateful to Simon Garrard for supplying this information.

[6]Wendy R. Childs and Maryanne Kowaleski, "The Internal and International Fish Trades of Medieval England and Wales," in Starkey, Reid and Ashcroft (eds.), *England's Sea Fisheries,* 29-35.

[7]Evan Jones, "England's Icelandic Fishery in the Early Modern Period," in *ibid.,* 105-110.

[8]Todd Gray, "Devon's Fisheries and Early-Stuart northern New England," in Michael Duffy, *et al.* (eds.), *The New Maritime History of Devon* (2 vols., London, 1992), I, 139-144.

from the late seventeenth century when the English government started to assemble statistical data in an increasingly systematic and comprehensive manner. To use these data as evidence with which to address historical problems, a number of questions have to be asked of the source material. In particular, it is important to establish: by whom, why, and by what procedures was the data originally generated? how reliable and representative is the information they yield? can it be corroborated? and how might the data be adapted to address specific issues, such as those at the core of the History of Marine Animal Population project?

Using the first of these questions to organize the material, the present survey assessed a range of qualitative and quantitative records relating to the output of the Newfoundland fishery. This appraisal entailed the examination of primary source materials generated by five types of author: those engaged in the fishing business; observers of the harvesting operation and/or the cod trade; pamphleteers keen to promote the business; state departments responsible for regulating maritime activity; and government agencies charged with administering the fishery. The first three forms of informant – practitioners, observers and pamphleteers – produced a proliferation of largely qualitative information. Collectively, this material offers revealing insights into the character of the Newfoundland trade. For instance, the journals of Richard Whitbourne, James Yonge and Joseph Banks provide first-hand documentary evidence of the prosecution of the fishery in the seventeenth and eighteenth centuries. Likewise, the business accounts and correspondence contained in the Fox Papers (Devon Record Office), the records of the Lester and Garland families (Dorset Record Office), and the archives of Newman, Hunt and Coave (Provincial Archives of Newfoundland) reveal much about the development of the trade in the pivotal decades of the late eighteenth and early nineteenth centuries. Opinions voiced in the petitions of the Western Adventurers, and testimony given at public hearings, such as that supplied by Peter Ougier of Dartmouth to the 1793 Enquiry into the State of the Newfoundland Trade, suggests how the merchants engaged in the fishery perceived its prospects and sought government assistance to alleviate its problems. Yet there are dangers here, for such evidence, like that of pamphleteers, was often partial in both senses of the word, as witnesses and advocates tended to present one-sided, incomplete versions of the truth. Such opinions are difficult to corroborate and should therefore be treated with caution, especially when they involve quantification. For example, it is often stated in historical works that upwards of 10,000 Englishmen sailed for Newfoundland at the start of fishing seasons in the early 1620s. This figure is probably too high.[9] Derived from a statement read to the House of Parliament by William Nyell, it almost

---

[9]Peter E. Pope, "Early Estimates: Assessment of Catches in the Newfoundland Cod Fishery, 1660-1690," in Daniel Vickers (ed.), *Marine Resources and Human Societies in the North Atlantic since 1500* (St. John's, 1995), 7-40.

certainly exaggerates the real magnitude of the fishery to support a case being presented to the chamber. Figures of 250 ships, 5000 men and 300,000 quintals of cod, furnished for the 1615 season in an isolated government return,[10] are perhaps more realistic, and certainly sit more comfortably with a recent estimate which suggests that 6000 men were engaged in the fishery during this period.[11]

For the sixteenth and seventeenth centuries, incidental, sometimes dubious data offers the only basis for gauging the scale of the harvesting operation in Newfoundland waters. In contrast, from the 1670s the progress of the fishery can be charted by virtue of a series of statistical return undertaken on behalf of government departments. These quantitative sources fall into two categories. First, there were the various accounts of shipping and seafarers drawn up by state departments responsible for administering the private maritime sector. Some of these yardsticks were driven by the long-held belief that the merchant shipping industry should be cultivated as a nursery of seafarers, whose skills, honed at private expense, should be transferred to the Navy in wartime.[12] While such motives underpinned the surveys that the Duke of Buckingham ordered in 1619 and 1621,[13] they were also evident in the assembly of trade and shipping figures from the mid-eighteenth century and in the ship registration scheme adopted in 1786. At the same time, these and other devices were deployed to regulate Britain's maritime services. The records of the High Court of Admiralty, for example, include a great many cases arising from disputes in the fishing industry,[14] while the welfare of seafarers and fishermen gave rise to the seamen's sixpenny records from 1747 onwards.[15]

---

[10]PRO, ADM 7/688.

[11]Peter E. Pope, "The South Avalon Planters, 1630 to 1700: Residence, Labour, Demand and Exchange in seventeenth-century Newfoundland" (Unpublished PhD thesis, Memorial University of Newfoundland, 1992), 207-208.

[12]David J. Starkey, "The West Country-Newfoundland Fishery and the Manning of the Royal Navy," in Robert Higham (ed.), *Security and Defence in South-West England before 1800* (Exeter, 1987), 93-101.

[13]Todd Gray, "The Duke of Buckingham's Survey of South Devon Mariners and Shipping, 1619," in Duffy, *et al.* (eds.), *New Maritime History of Devon*, I, 117-118.

[14]John C. Appleby and David J. Starkey, "The Records of the High Court of Admiralty as a Source for Maritime Historians," in David J. Starkey (ed.), *Sources for a New Maritime History of Devon* (Exeter, 1986), 70-85.

[15]Ralph Davis, "Seamen's Sixpences: An Index of Commercial Activity, 1697-1828," *Economica*, XXIII (1956), 328-343.

The value of such material to studies of fishing output lies in the contextual evidence it supplies and the corroboration it provides for more specific data on the fisheries. In particular, it offers a check on sources relating to fishing effort. In the 1619 survey of mariners, for instance, it was recorded that 279 vessels belonged to the South Devon ports of Plymouth, Exeter, Dartmouth and Torbay.[16] As this region was at the heart of the Newfoundland trade, though it was by no means the only area to participate, a migratory fleet of 250 vessels, as indicated in the return of 1615, would seem to be a reasonable estimate. Likewise, the muster rolls collected to account for the sixpences taken out of each seafarer's monthly wage can be used to gauge the origins and destinations of vessels sailing into and out of British ports. As the Dartmouth muster rolls date back to 1772, and Exeter's survive from 1800, these data facilitate analysis of a large proportion of the vessels and crews engaged in the Newfoundland fishery.[17] In effect, this provides a proxy measure of the fishing effort – as well as the patterns of migration[18] – of the South Devon heartland of England's Newfoundland interests.

The second form of quantitative data was collected by fishing "admirals" and naval officers working at Newfoundland and returned to the Lords Commissioners for Plantations and Colonies. Providing information on the output, fishing effort, prices, population and other facets of English enterprise at Newfoundland, these data are available in a comparatively rudimentary form for eleven of the years from 1675 to 1697.[19] Then they survive more fully for most years from 1698 to 1833.[20] The series continues from 1833 to 1903,[21] though it is not entirely compatible with the earlier series as the output figures constitute approximations based on the quantity of dried cod exported each year, augmented by an estimate of the amount consumed on the island. Other problems arise in using the material. For instance, it is difficult to determine the accuracy of the information provided, though it appears to be internally consistent and is broadly corroborated by qualitative evidence. In particular, the depression in the output figures evident during the 1710-1739 period (see appendix table 1) tallies well with a welter of

---

[16]Gray, "Duke of Buckingham's Survey," 117-118.

[17]PRO, BT 98/3-9 (Dartmouth); and BT 167/39-41 (Exeter).

[18]See W. Gordon Handcock, *Soe Longe as There Comes no Women: Origins of English Settlement in Newfoundland* (St. John's, 1989).

[19]PRO, BT 6/57.

[20]PRO, CO 194; and Shannon Ryan, "An Abstract of CO 194 Statistics" (Unpublished manuscript, Memorial University of Newfoundland, 1969).

[21]PRO, CO 199.

complaints and petitions emanating from the West Country fishing merchants at that time. The coverage of the material warrants further investigation of the kind applied to the 1675-1698 returns.[22] An intensive search of PRO, CO 1 holdings, for instance, might fill the evidential gaps that appear the CO 194 series for the 1746-1770 period, when various fishing districts – such as Trepassey, Renewse, Placentia and Old Perlican – were unreported. Nevertheless, the data provided by the colonial office returns (CO 194 and CO 199) are the most important source of statistical information on the British Newfoundland fishery in the eighteenth and nineteenth centuries.

A more serious shortcoming of the British statistics, in terms of the Newfoundland fishery as a whole, is that they contain little material on the output of foreign fishermen. For instance, in the seventeenth century, few references to the scale of French fishing have been uncovered in English repositories, the most widely known being the naval return of 1677 which indicates that the French deployed 102 ships and 1836 boats in the fishery, an effort that yielded an estimated 550,800 quintals of dried cod.[23] This output was more than double the estimated English return of 238,000 quintals.[24] During the eighteenth century, the French catch generally exceeded that of the British until the Seven Years' War (1756-63) when the tables were turned and remained so for much of the next three decades.[25] This is evident in the French production figures for the years 1769-1774 and 1786-1792 that were printed in the "Second Report of the Enquiry into the State of the Newfoundland Trade."[26] Turgeon's estimates concur with the data for the earlier period, but diverge considerably for the years from 1786 to 1792, when the "British admirals upon that coast" returned figures less than half of those derived from French sources.[27]

The British statistical returns have been used to chart the changing configuration of the fishery, notably the emergence of the bye-boat and banker modes of production, and the shift from migratory to sedentary in the operational

---

[22]See Pope, "Early Estimates."

[23]PRO, ADM 7/369.

[24]Pope, "Early Estimates," 24.

[25]Laurier Turgeon, "Fluctuations in Cod and Whale Stocks in the North Atlantic during the Eighteenth Century," in Vickers (ed.), *Marine Resources and Human Societies*, 87-120.

[26]See Sheila Lambert (ed.), *House of Commons Sessional Papers of the Eighteenth Century* (Wilmington, DE, 1975), XC.

[27]Turgeon, "Fluctuations," 106.

base of the fishing operation.[28] Moreover, these detailed figures have been deployed to highlight the impact that the regular bouts of European colonial warfare had on the output and conduct of the fishery.[29] However, the data also offer evidence to sustain analyses of the scale of fish harvesting in Newfoundland waters over the long run, and to explore the relationship between fishing effort, fish production and fish stocks.

Tentative Findings

This survey has found no reason to challenge the conventional wisdom that the English were relatively minor participants in the Newfoundland fishery until the late sixteenth century. Nor has it found cause to disagree with the contention that Englishmen, based chiefly in the West Country, expanded their distant-water fishing activities considerably during the 1590-1620 period, before struggling in the mid-century decades. However, the evidential basis of these assertions remains somewhat fragile. More detailed enquiries into the holdings of English archives – such as those undertaken by Childs and Kowaleski on the medieval period, Jones on the Iceland trade, and Pope on the late seventeenth century – is therefore required to confirm or adjust the general picture that derives from qualitative and circumstantial evidence.

The existence of a substantial core of statistical data for the post-1675 era offers a clearer view of the scale of the English Newfoundland fishery. Analysis of these data suggests that the output of the fishery in the eighteenth century consistently exceeded the estimated figures offered by Hutchings and Myers,[30] as indeed seems to be the case for the late seventeenth century.[31] As table 1 shows, the English fishery alone generally exceeded 100,000 tons from the 1750s onwards, sometimes, as in 1788 and consistently from the 1820s, by a consider-

---

[28]Keith R. Matthews, "A History of the West of England-Newfoundland Fisheries," (Unpublished DPhil thesis, University of Oxford, 1968).

[29]David J. Starkey, "Devonians and the Newfoundland Trade," in Duffy *et al.* (eds.), *New Maritime History of Devon*, I, 163-171.

[30]Jeffrey A. Hutchings and Ransom A. Myers, "The Biological Collapse of Atlantic Cod off Newfoundland and Labrador: An Exploration of Historical Changes in Exploitation, Harvesting Technology, and Management," in Ragnar Arneson and Lawrence Felt (eds.), *The North Atlantic Fisheries: Successes, Failures and Challenges* (Charlottetown, PEI, 1995), 39-93.

[31]Pope, "Early Estimates;" and Jeffrey A. Hutchings, "Spatial and Temporal Variation in the Exploitation of Northern Cod, *Gadus morhua*: A Historical Perspective from 1500 to Present," in Vickers (ed.), *Marine Resources and Human Societies*, 41-68.

able margin. Adding the French output to these totals for the 1769-1774 period, the total live weight catch ranged between 204,000 and 275,000 metric tons, indicating that more cod was harvested earlier than previous work has suggested.[32]

Table 1 further points to the vacillations that marked the annual catch of the English fishery from the late seventeenth century, albeit around a steadily rising trend. Ranging in the eighteenth century from a nadir of 100,823 quintals to a peak of 948,970 quintals, the fishery was prone to fluctuate. The limited degree to which this was due to variations in the scale of the fishing effort is implied in appendix table 2. This deploys the methodology devised by Walter Garstang to test the impact of human predation on North Sea stocks in the 1890s,[33] since adapted by Pope to analyse fishing effort and output in Newfoundland, 1675-1698. Accordingly, the rate at which quintals of dried fish were taken by the boats employed in the fishery – by inhabitants, British fishing ships and bye-boatkeepers – is presented. The results suggest that productivity was generally much higher than the standard of 200 quintals per boat set in the seemingly "good years" of the early seventeenth century.[34] Indeed, it was only in the 1710-1730 period that the rate remained consistently and significantly below par. That this was an era of depression in the fishery, as signalled by qualitative evidence and output figures, indicates that factors beyond the control of human predators were at work.

Such findings are essentially preliminary. The statistical series on which they are based require further validation and refinement. Moreover, as with the pre-1675 era, there is a need for further research into the vast archives generated by British departments of state – those of the Colonial Office, Board of Trade, and Chancery are the obvious candidates – to locate information that would supplement, corroborate or supplant the data that has already surfaced. While the success of such a trawl is by no means guaranteed, it would be very surprising if it did not yield some substantial documentary catches.

---

[32]Hutchings and Myers, "Biological Collapse," figure 2.

[33]Micahel S. Haines, "Britain's Distant Water Fishing Industry, 1830-1914: A Study in Technological Change" (Unpublished PhD Thesis, University of Hull, 1998).

[34]Pope, "Early Estimates."

Table 1
Fish Catches in the Newfoundland Fishery,
Selected Years, 1698-1899
Dry quintals (metric tonnes live weight)

	British Catch	French Catch
1698	272618 (65428)	
1708	135934 (32624)	
1713	103750 (24900)	
1718	100823 (24198)	
1723	139756 (33541)	
1729	170220 (40853)	
1738	350979 (84235)	
1748	483800 (116112)	
1758	374710 (89930)	
1768	573450 (137628)	
1769	578024 (138726)	275000 (66000)
1770	649498 (155880)	435340 (104482)
1771	644919 (154781)	310000 (74400)
1772	759843 (182362)	388800 (93312)
1773	780328 (187279)	336250 (80700)
1774	695866 (167008)	386215 (92692)
1778	386530 (92767)	
1788	948970 (227753)	
1797	374970 (89993)	
1808	478735 (114896)	
1818	606733 (145616)	
1828	933062 (223935)	
1838	824515 (197884)	
1849	1345167 (322840)	
1859	1305793 (313390)	
1869	1074106 (257785)	
1879	1236395 (296735)	
1889	1185576 (284538)	
1899	1529396 (367055)	

Note: Conversion rate: 1 quintal = 0.24 metric tonnes live weight fish. See Pope, "Early Estimates," note 7.

Sources: 1698-1828 (British): PRO, CO 194; *1838-1899*: approximations from PRO, CO 199; *1769-1774 (French)*: Second Report of the Enquiry into the State of the Newfoundland Trade, 1793, printed in Sheila Lambert (ed.), *House of Commons Sessional Papers of the Eighteenth Century* (Wilmington, DE, 1975), XC.

Table 2
Fish Catch and Catch Rates, British Newfoundland Fishery,
Selected Years, 1675-1828
Catch in dry quintals. Rate in quintals/boat

	Catch	Boats	Rate
1676-1681	221,000	1190	185.7
1692-1698	306,000	960	318.8
1698	272,618	929	293.5
1708	135,934	536	253.6
1713	103,750	645	160.9
1718	100,823	904	111.5
1723	139,756	966	144.7
1729	170,220	685	248.5
1738	350,979	1126	311.7
1748	483,800	1226	394.6
1758	374,710	1315	285
1768	573,450	2104	272.6
1778	386,530	1701	227.2
1788	948,970	2680	354.1
1797	374,970	1263	396.9
1808	478,735	1778	269.3
1818	606,733	3162	191.9
1828	933,062	3160	295.3

Sources: 1676-1681, 1692 and 1698: Peter E. Pope, "Early Estimates: Assessment of Catches in the Newfoundland Cod Fishery, 1660-1690," in Daniel Vickers (ed.), *Marine Resources and Human Societies in the North Atlantic since 1500* (St. John's, 1995), 7-40; 1698-1828: PRO, CO 194.

Testing Ecological Models:
The Influence of Catch Rates on Settlement
of Fishermen in Newfoundland, 1710-1833[1]

Ransom A. Myers

Abstract

Perhaps the most fundamental theory of natural renewable resource exploitation is Scott Gordon's bio-economic model for an open access fishery. I test this theory using a superb data set from fishing communities in three bays in Newfoundland from 1710 to 1833. Gordon's theory predicts very well the pattern of settlement in these three bays during this period. Fishermen in each tended to settle when catch rates were greater than forty quintals of dried salt cod (about ten metric tonnes of fresh cod) per man a year to emigrate when catch rates were less. The mean catch rate per man per year during the 1700s and early 1800s remained remarkably constant, presumably because of this population movement. Remarkably, the catch rate of ten metric tonnes per year was the average of inshore fishermen in Newfoundland until the collapse of the fishery in the 1990s.

Introduction

In a seminal analysis, Scott Gordon developed a simple, yet profound, model of why natural renewable resource exploitation to become unprofitable over time.[2] He demonstrated that new exploiters entered the system until the profit margin reached zero. Although there is wide acceptance of Gordon's model, there are few analyses that firmly demonstrate its empirical validity.[3]

---

[1]I would like to thank Daryl Janes, Stacy Fowlow, and Nicholas Barrowman for assistance. Suggestions by Jeff Hutchings, Tim Smith and David Starkey improved the manuscript.

[2]H.S. Gordon, "Economic Theory of a Common-Property Resource: The Fishery," *Journal of Political Economy*, LXII (1954), 124-142.

[3]See, for example, C.W. Clark, *Mathematical Bioeconomics: The Optimal Management of Renewable Resources* (2nd ed., New York, 1990).

The Gordon model of the bioeconomics of the exploitation of fish populations forms the basis for many of our present attempts to understand and regulate commercial fisheries.[4] The model, and its subsequent elaborations, describes the entry of fishermen into the fishery in terms of the profit or catch rates. The purpose of this essay is to examine empirically a key component of this model: the behaviour of fishermen in entering and leaving a fishery, using an extraordinary time series of population and catch data for the Newfoundland salt cod fishery from 1710 to 1833. There are several reasons that Newfoundland in this period offers an ideal case to test the theory. First, excellent records exist. There are annual reports that specify details of the fishery and settlement for nearly all years. Second, there was an almost unlimited opportunity for settlement. There were hundreds of thousands of migrant fishermen in the region, but the total resident population was less than 20,000 by the 1790s.[5] A further advantage of these data is that it is possible to break them down into distinct regions, enabling me to cross-validate the models and parameter estimates of "replicate realizations" of the same process.

During this period the shore-based cod fishery along the east and north coasts of the island of Newfoundland was exclusively in the hands of the British. The only other cod during this period was an offshore fishery by the British and French on the Grand Banks. These cod are now believed to be largely an independent stock from those nearer to shore.[6] The fishery in each harbour in the British fishery was monitored by a "fishing admiral," who was required to report each year the catch, number of fishermen, and a variety of demographic and economic information. Although there are gaps and various problems with these data, they allow us to study entry into the fishery, in particular when fishermen entered and left. We compiled what appeared to be reliable data from three regions, a procedure that allowed our test to be replicated: Trinity Bay, Conception Bay, and the region from St. John's to Cape Race (see figure 1). In this article I will concentrate on testing Gordon's bioeconomic theory to explain settlement. I will also address other hypotheses briefly in the section headed "discussion."

---

[4]C.W. Clark, *Bio-economic Modelling and Fisheries Management* (New York, 1985).

[5]W.G. Handcock, *Soe Longe as There Comes Noe Women: Origins of English Settlement in Newfoundland* (St. John's, 1989).

[6]D.E. Ruzzante, C.T. Taggart and D. Cook, "A Nuclear DNA Basis for Shelf- and Bank-Scale Population Structure in Northwest Atlantic Cod (*Gadus morhua*): Labrador to Georges Bank," *Molecular Ecology*, VII, No. 12 (December 1998), 1663-1680.

Figure 1
Trinity and Conception Bays; St. John's to Cape Race;
and Modern Fishing Regions

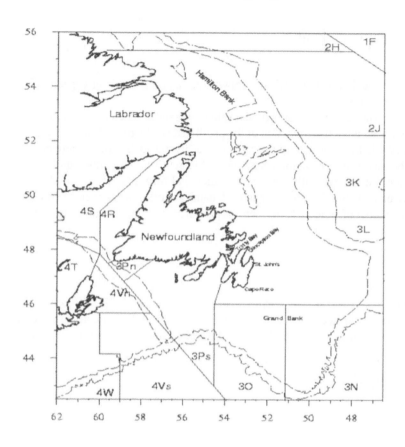

Source: See text.

Gordon's Bioeconomic Model

My goal is to describe empirically the relationship of the change in the number of fishermen to the catch per unit effort (CPUE). I will use a dynamic version of Gordon's open-access fishery model developed by Barry Smith.[7] Consider the dynamics of a fish population, X, given by the Shaffer logistic model, where

---

[7]J.B. Smith, "Stochastic Steady-State Replenishable Resource Management Policies," *Marine Resource Economics*, III , No. 2 (1986),155-168.

harvesting is a simple function of the ability to catch fish, q, and effort, E. That is,

$$\frac{dX}{dt} = rX\left(1 - \frac{X}{K}\right) - qEX,$$

where r and K are the intrinsic growth rate and the carrying capacity of the population, respectively.

The key to Gordon's model is that the rate at which fishermen enter the fishery is an increasing function of the profit that can be made from fishing. That is, they join the fishery until the catch rate decreases to the point where no profit can be made: the population may be said to reach a "bionomic equilibrium." Gordon did not explicitly specify the functional form for the rate of change of fishermen, but Smith assumed that the dynamics of effort – in our case, the change in the number of resident fishermen – is proportional to current revenues. That is,

$$\frac{dE}{dt} = k\pi,$$

where k is a constant. The current net revenues are simply the difference in the revenues – the price of the catch, p, times the catch – which are assumed to be a linear function of effort, qEX, minus the cost of one unit of effort, c. That is,

$$\frac{dE}{dt} = k(pqX - c)E,$$

or

$$\frac{d(\log E)}{dt} = k(pqX - c).$$

Since we do not know the true abundance of fish, the above equation will be examined in terms of the CPUE, which is assumed to be proportional to true abundance, i.e., C/E = qX. Thus, we have

$$\frac{d(\log E)}{dt} = k\left(p\frac{C}{E} - c\right).$$

This formulation is for simplicity; a more complex model developed by Clark shows that investment decisions may be better described as a threshold: fishermen enter the fishery if it is profitable to do so.[8] The point I wish to make here is that the derivative of the log of effort – for example, the number of fishermen – should be an increasing function of CPUE. This is at the heart of most models of fishermen's behaviour, and it is this relationship that I will test.

Data on the Salt Cod Fishery in Newfoundland

Data were compiled annually throughout the eighteenth and into the nineteenth century by the Colonial Office in London on the fishery in each major harbour in Newfoundland. During this period, fishermen on the east and north coasts of Newfoundland were mainly from the "west country" of England, especially Devon and Cornwall, or southern Ireland. There were three categories of fishermen: migrants, "bye boatmen," and inhabitants. While migrants transported their fishing boats to Newfoundland each year, bye boatmen kept their boats in Newfoundland harbours and migrated to the island annually for the summer fishery. Inhabitants settled on the island. I will concentrate on the dynamics of the inhabitants because they had to be committed to the fishery to sustain a permanent home in Newfoundland. It is also clear that they had greater access to the best fishing areas.[9] In many ways this is a perfect empirical situation to test a model of open-access entry to a fishery because high mobility was feasible. It was relatively easy to settle after becoming established in the fishery. The land-based fishery was carried out from small boats, often shallops, using baited hooks. The cod were salted and dried on land.

There have been several studies using the data from Colonial Office (CO) 194.[10] I started with a compilation of the Forsey and Lear material, but carried out extensive checks using the raw data in the microfilmed edition of the CO 194 statistics at the Centre for Newfoundland Studies at Memorial University of Newfoundland. The records recorded the statistics by fishing harbour for each year. The initial checking consisted of three steps: a comparison of the data from the three studies cited in note 10; a check of the microfilm records to determine if data from any fishing harbours were missing for a given year; and a check of

---

[8]See Clark, *Mathematical Bioeconomics*.

[9]C.G. Head, *Eighteenth Century Newfoundland: A Geographer's Perspective* (Toronto, 1976).

[10]*Ibid.*, R. Forsey and W.H. Lear, "Historic Catches and Catch Rates of Atlantic cod at Newfoundland during 1677-1833," *Canadian Data Report on Fisheries and Aquatic Sciences*, DCLXII (1987); and S. Ryan, *Fish Out of Water: The Newfoundland Saltfish Trade 1814-1914* (St. John's, 1986).

any sudden fluctuations in any of the time series. There were numerous mistakes detected in the Forsey and Lear study, including addition errors, totals for regions where data from some harbours were missing, and highly unlikely fluctuations from year to year in catch, effort, or number of inhabitants. These could almost always be resolved by examining the original census reports. For some years data were missing for a harbour in one of the regions I considered. In such cases the data were excluded for the entire region for that year.

Figure 2
Men Per Boat over Time for Trinity and Conception Bays,
and St. John's to Cape Race

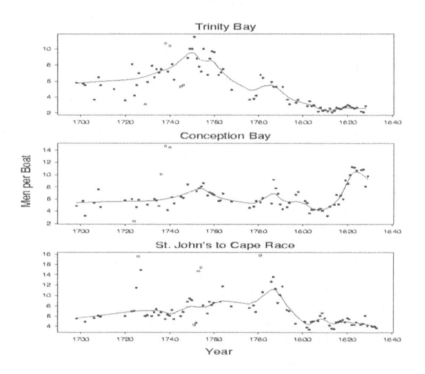

Note: The supersmoother technique was used (solid line) and outliers are shown as open circles. The outliers for men per boat were determined using the supersmoother and considered to be greater than two or less that -2 standard residuals, where standard residual was calculated to be the residual/standard error.

Source: See text.

I further eliminated outliers by examining the trend over time in the number of men per boat, since it was unlikely that crew size would change abruptly. A crew typically consisted of three fishermen per boat, plus two men processing on shore.[11] Few women were involved in the fishery during this period.[12] But the number of men per boat changed over time and varied between regions (see figure 2). Some of the sudden shifts in the number of men per boat were unlikely. For example, in Conception Bay the average number of men per boat increased from 4.2 in 1739 to 14.5 in 1740. Since it is impossible to understand why this might have occurred, I eliminated all data where the standardized residual from a long-term smooth curve had an absolute value greater than two. The long-term mean was defined by a "supersmoother," and the standardized residuals were calculated by dividing by the standard error from the smooth.[13] The long-term mean men per boat varied between three and seven. This eliminated between four and six years of data per region. While the causes of the long-term changes in the number of men per boat are unclear, it is important that any analysis should be sensitive to them. For this reason, I will here consider two indices of CPUE, catch per boat and catch per man.

Although some data go as far back as 1677, I started in 1710 because they were collected more consistently thereafter; indeed, the years 1710-1833 were remarkably homogeneous in this regard. The last yearly census was in 1833.[14] Although there were numerous wars in this period, there were no French raids on English settlements in Newfoundland. Nor were other nations allowed to fish in any of the three regions in question.

Catch is measured in "quintals," 112 pounds of split, salted, and dried cod. It took about one metric ton of fresh cod to produce 4.2 quintals of dried fish. The derivative, $\frac{d(\log E)}{dt}$, was estimated by smoothing logE and calculating the finite difference of the smoothed curve one-half year before and after the year of interest. Smoothing was necessary because of the large variation in the estimated

---

[11]Head, *Eighteenth Century Newfoundland*.

[12]Handcock, *Soe Longe as There Comes Noe Women*.

[13]This procedure is discussed in J.H. Friedman, "A Variable Span Smoother," Stanford University, Department of Statistics, Laboratory for Computational Statistics, Technical Report No. 5, 1984.

[14]S. Ryan, *Abstract of Returns for the Newfoundland Fishery 1698-1833* (St. John's, 1969).

number of fishermen. A robust local smoother (lowess), with a window of fifteen years, was used for this smoothing.[15]

It is not a simple matter to assess the statistical significance of the correlation between $\frac{d(\log E)}{dt}$ and CPUE because of auto-correlation and a large number of missing values. A "Monte Carlo" approach was used to estimate the probability of the observations occurring by chance.[16] I modelled the $\frac{d(\log E)}{dt}$ and CPUE as first-order auto-regressive processes, with the auto-correlation estimated from each series, and generated 1000 pairs of time series with the characteristics and length of the observed ones (except there were no correlation between them). Data points in the simulated time series were removed in the same pattern as the observed ones, and histograms of the correlations between the simulated time series were plotted. The statistical significance of an observed correlation was estimated from these histograms.

An alternative approach to estimate statistical significance is to use a product-moment correlation and to correct the number of degrees of freedom by the degree of auto-correlation in the two time series. This will effectively reduce the degrees of freedom for tests of statistical significance.[17] Since both methods gave similar results, I will only report the results from the Monte Carlo method.

Results

The basic data used in the analysis were the number of men, catch per boat and catch per man (figures 3-5). While catch rates in 1710 were typical of those of later times, the catch rates between 1714 and 1727 were consistently the lowest in the time series. Keith Matthews has detailed this collapse of the fishery; it was perhaps the worst period before the collapse in the 1990s, and there was virtually no new settlement during this period.[18]

Higher catch rates were observed around 1745 for all regions; this increase corresponds to the largest relative growth in population during the time period. In Trinity Bay, the population increased from around 400 men in the early

---

[15]See W.S. Cleveland, "Robust Locally Weighted Regression and Smoothing Scatterplots," *Journal of the American Statistical Association*, LXXIV (1979), 829-836.

[16]B.F.J. Manly, *Randomization and Monte Carlo Methods in Biology* (New York, 1991).

[17]M.B. Priestley, *Spectral analysis and Time Series* (2 vols., London, 1981), I.

[18]K. Matthews, "A History of the West of England-Newfoundland Fishery" (Unpublished DPhil thesis, Oxford University, 1968).

1700s to over 1000 when the catch rates increased around 1740. Similar increases in catch rates and population size occurred in Conception Bay and St. John's to Cape Race. After 1750, catch rates fell to around 200 to 250 quintals per boat, and there was very little change in population for at least thirty years.

Figure 3
Number of Men, Catch Per Boat (Quintals/Boat) and
Catch Per Man (Quintals/Man) for Trinity Bay

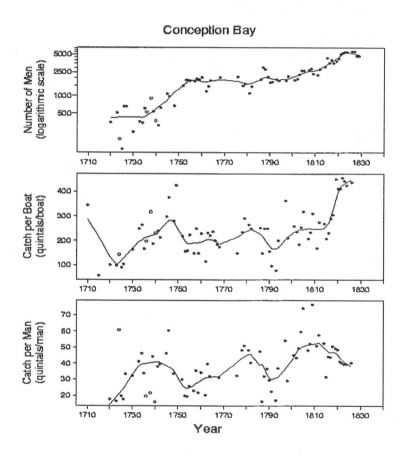

Note: A lowess smoother with a window of fifteen years was used (solid line) and outliers in men per boat (figure 2) are shown as open circles.

Source: See text.

Ransom A. Myers

Figure 4
Number of Men, Catch Per Boat (Quintals/Boat) and
Catch Per Man (Quintals/Man) for Conception Bay

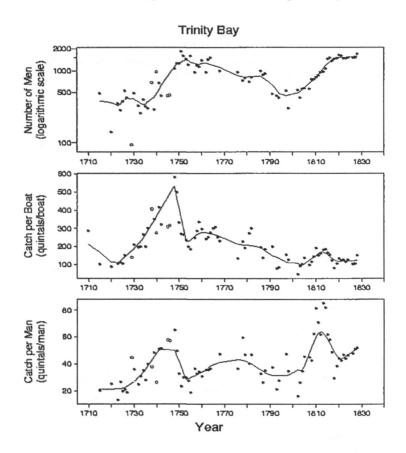

Note: See figure 3.

Source: See text.

After this time, however, there were significant differences among the regions. In Trinity Bay, population and catch per boat declined to 1800. Then about 1810, catch per boat, catch per man, and number of settlers increased. After 1820, the time series stabilized. In Trinity Bay, the correlation of the population growth rate with catch per man and catch per boat was 0.61 (p=0.022) and 0.48 (p=0.38), respectively.

In Conception Bay, there was a relatively constant increase in population from 1800 to 1820, which corresponded to a high catch rate per man. The catch per boat was about average until 1820, when it suddenly increased. In Conception Bay, the correlation of population growth rate with catch per man and catch per boat was 0.30 (p=0.32) and 0.31 (p=0.32), respectively.

Figure 5
Number of Men, Catch Per Boat (Quintals/Boat) and
Catch Per Man (Quintals/Man) for St. John's to Cape Race

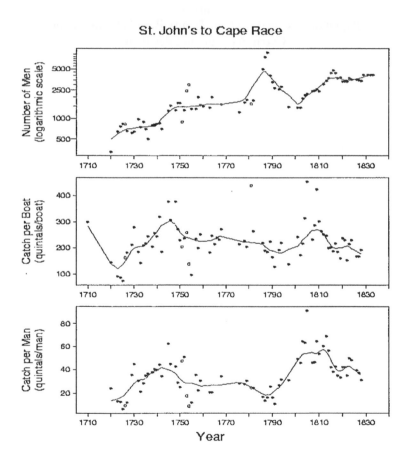

Note: See figure 3.

Source: See text.

From St. John's to Cape Race, the number of men increased suddenly around 1785, but this did not correspond to an increase in either catch rate. The decline in population from 1785 to 1800 corresponded to low catch rates, while the increase from 1800 to 1815 corresponded to high catch rates. The deviation around 1788 appears to have been caused by an overestimation of the number of fishermen in the source, but I was unable to verify this from the microfilm records. After 1815, the population and catch rates stabilized. From St. John's to Cape Race, the correlation of population growth rate with catch per man and catch per boat was 0.48 (p=0.107) and 0.46 (p=0.114), respectively.

Figure 6
Population Growth Rate in Trinity Bay, 1710-1833, and
Catch Per Man and Per Boat

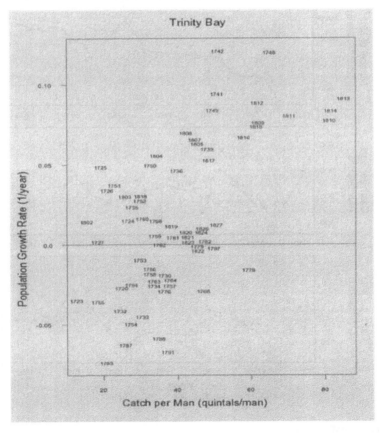

Note: A reference line at zero is given (dotted line).

Source: See text.

In all three regions, the population growth rate, $\frac{d(\log E)}{dt}$, was an increasing function of CPUE as measured by catch per man or per boat. The positive relationship was strongest for Trinity Bay (figure 6) and the St. John's to Cape Race region (figure 8). The relationship was still positive, but much less so, for Conception Bay (figure 7). The statistical significance of the results was low despite the relatively long time series because of strong auto-correlation in the data: there are relatively few true degrees of freedom. Nevertheless, the relationships are surprisingly consistent among regions.

Figure 7
Population Growth Rate in Conception Bay, 1710-1833, and Catch Per Man and Per Boat

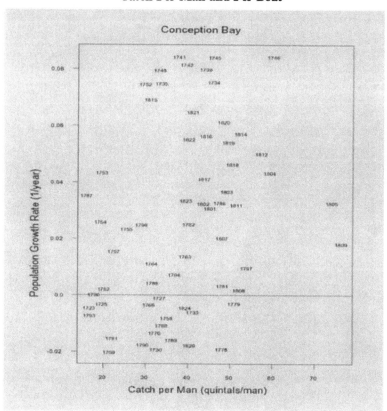

Note: See figure 6.

Source: See text.

I repeated the analysis using various modifications to test the robustness of the results. Different widths for the "window" were used to estimate the population growth rate. A width of ten or twenty years resulted in only a small modification. The analysis was also carried out without removing the outliers; the correlations between population growth rate and catch rate were reduced only slightly. In short, the results appear to be robust.

Figure 8
Population Growth Rate in St. John's to Cape Race, 1710-1833, and Catch Per Man and Per Boat

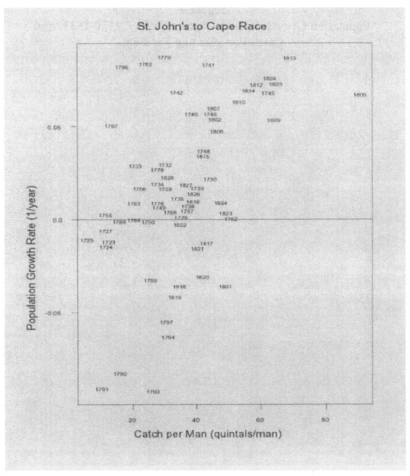

Note: See figure 6.

Source: See text.

A Comparison of Modern and Historic Catch Rates

The long-term catch in each region was about forty quintals per man, or about ten metric tonnes of fresh cod per year. This was also the approximate level where the derivative of population growth rate was positive when plotted against catch rate. That is, when the catch rate was above this level, on average the population increased, while when it was below this level population decreased.

The catch rate of about ten metric tonnes per person appears to be typical also of the catch rates in the seventeenth-century British fishery in Newfoundland and the French fishery in Placentia and Isle Royale (present-day Cape Breton).[19] (Turgeon 1995). Remarkably, they are also typical of late twentieth-century Newfoundland catch rates. For example, in the 1980s the cod catch around Newfoundland was around 200,000 metric tonnes, while the number of cod fishermen was approximately 20,000. That is, catch per man in Newfoundland appears to have been the same for the last four centuries.

Discussion

It appears that the settlement of fishermen in Newfoundland from 1710 to 1833 can largely be explained by the CPUE. When catch rates were good, fishermen settled. While other factors, such as wars, colonial settlement policies and prices, may have affected settlement, the rational choice to settle in Newfoundland would have been made only if catch rates were good. There was little incentive to overwinter if they were poor.

It is not clear if wars had a large impact on settlement. For example, David Starkey has pointed out that the ratio of migratory to resident fishermen decreased during each war between 1698 to 1826.[20] But conflicts may not have been related as directly to the decision to settle as much as the ability to take part in the migratory fishery. Even during wars, many migrant fishermen came to Newfoundland, and they could have settled had they wanted to do so.

The results for the three study areas are consistent with the behaviour predicted by Gordon's model. This was true for alternative estimates of CPUE and alternative treatments of outliers, with consideration of reasonable lags. The relationship was strongest for Trinity Bay and the St. John's to Cape Race region,

---

[19]On the British fishery, see P. Pope, "Early Estimates: Assessment of Catches in the Newfoundland Cod Fishery, 1660-1690," in D. Vickers (ed.), *Marine Resources and Human Societies in the North Atlantic since 1500* (St. John's, 1995), 9-40. For the French experience, see L. Turgeon "Fluctuations in Cod and Whale Stocks in the North Atlantic during the Eighteenth Century," *ibid.*, 89-121.

[20]D.J. Starkey, "Devonians and the Newfoundland Trade," in M. Duffy, *et al.* (eds.), *The New Maritime History of Devon* (2 vols., London, 1992), I, 163-171.

and weakest for Conception Bay. For all regions, settlement did not usually occur unless catch rates were good – around forty quintals of salted, dried cod per man per year, or around ten metric tonnes of fresh cod harvested. (ten tonnes of cod harvested per year per fisherman is the approximate average catch for the last 400 years.) These results appear to be robust and held up when alternative methods were used.

There are two general explanations as to why the CPUE changed over time. First, CPUE reflects changes in fish abundance or distribution caused by environmental changes. While environmental changes certainly influence cod populations, this is unlikely to be the full answer. The second general explanation is that fishing affected cod abundance as early as the 1700s.[21] The catch in the British portion of eastern Newfoundland possibly was as high as 100,000 tonnes by the 1600s, a level that could certainly have caused the collapse of local cod populations.[22] Analyses of historic tagging data have shown that the inshore fishery since 1945 was capable of causing very high mortality rates in fish stocks.[23] There is direct evidence of overfishing in the historic record as well. Laurier Turgeon has shown that the intensive British and French inshore fisheries in the eighteenth century were associated with a decline in the ratio of cod liver oil to dried fish produced, which indicates a decline in the average size of fish in the catch.[24] The analysis of catch rates presented here is consistent with the hypothesis that the inshore fishery was capable of causing very high fish mortalities.

One key factor may be changes in "catchability," q, over time. As fishing technology changed, catch rates may have improved.[25] This would have led to increased settlement until the catch rates returned to their previous levels.

---

[21]Head, *Eighteenth Century Newfoundland*.

[22]Pope, "Early Estimates."

[23]R.A. Myers, *et al.*, "The Collapse of Cod in Eastern Canada: The Evidence from Tagging Data," *ICES Journal of Marine Sciences*, LIII, No. 3 (1996), 629-640; and R.A. Myers, N.J. Barrowman and J.A. Hutchings, "Inshore Exploitation of Newfoundland Atlantic Cod (*Gadus morhua*) since 1948 as Estimated from Mark-Recapture Data," *Canadian Journal of Fisheries and Aquatic Sciences*, LIV, Supplement 1 (1997), 224-235.

[24]Turgeon, "Fluctuations."

[25]J.A. Hutchings and R.A. Myers, "The Biological Collapse of Atlantic Cod off Newfoundland and Labrador: An Exploration of Historical Changes in Exploitation, Harvesting Technology and Management," in R. Arnason and L. Felt (eds.), *The North Atlantic Fisheries: Successes, Failures and Challenges* (Charlottetown, 1995), 38-93.

One of the most surprising results of this analysis is that the fishery appears to have been close to bionomic equilibrium over the entire time period. The only time the fishery was not in this state was during periods of poor catches, such as the years around 1720, or when the time lag required for immigration created a temporary disequilibria. That is, the fishery appears to have been fully exploited for the type of gear and fishery practices used at the time, and by some definitions even over-exploited; from 1710 to 1833.

Open Questions

This paper only touches the surface of the vast archives of the colonial powers. Jeffrey Hutchings and I have compared environmental records with landings in Newfoundland from 1800 to 1995 and found no obvious link.[26] This analysis shows clear periods of low catch per unit effort in the inshore fishery; it would be interesting, given the slow recovery of cod stocks in the region from overfishing, to know if these periods of poor recruitment were associated with poor environmental conditions.

I have not included information on prices in this analysis. But a more sophisticated analysis should include such factors. A question of keen interest to fisheries ecologists is the total catch removed from the ecosystem. While Hutchings and I attempted to compile known estimates, most such analyses are badly flawed because they take no account of unreported catch. Peter Pope obtained much superior estimates for the period 1675-1698 using a ratio estimator, that is, by multiplying the number of fishermen by an average catch rate. His analysis shows that the catch per man fluctuated around forty quintals per man per year, with immigration occurring when catch rates were above that level and emigration occurring when they were below.[27] Thus, in the long term the catch man has been relatively constant. This general result may allow much better estimates of total catches to be estimated using modern mixed-effect and hierarchical Bayes methods.

---

[26]J.A. Hutchings and R.A. Myers, "What Can Be Learned from the Collapse of a Renewable Resource? Atlantic Cod, *Gadus morhua*, of Newfoundland and Labrador," *Canadian Journal of Fisheries and Aquatic Sciences*, LI, No. 9 (1994), 2126-2146.

[27]Pope, "Early Estimates."

Nineteenth-Century Expansion of the Newfoundland Fishery for Atlantic Cod: An Exploration of Underlying Causes

Sean T. Cadigan and Jeffrey A. Hutchings

Abstract

Few explanations have been offered for the northward expansion of the Newfoundland fishery for Atlantic Cod (Gadus morhua) into the waters off the Labrador coast. Extant literature alludes to this expansion as part of a British strategy to reinforce imperial control over coastal Labrador, or as a result of the need to deploy the otherwise unused capacity of schooners used in the spring seal hunt. A review of contemporary press accounts suggests that ecological problems in the inshore cod fishery of Newfoundland are the most important explanation of the development of the cod fishery in Labrador waters. A study of these accounts, combined with analysis of data from the Newfoundland censuses between 1845 and 1911, suggest the Labrador fishery expanded as catches and catch rates declined in the inshore cod fishery, despite increased fishing effort and increased harvesting capacity. We suggest that demographic growth and increased capital investment led to an ecological imbalance between Newfoundland fishing communities and the sub-stocks of northern cod upon which they had relied for their livelihoods. The colonial state and fish merchants responded with subsidy programs and credit strategies that encouraged members of the fishing communities of Newfoundland's northeast coast to invest in larger, schooner-rigged fishing vessels. The purpose of these larger vessels was to allow Newfoundland fishers to search further northward in search of new sources of cod.

Introduction

The Newfoundland fishery for Atlantic cod, *Gadus morhua*, was once the largest and the most productive cod fishery in the world.[1] From the late fifteenth century until the early 1990s, the "northern cod" component of this fishery constituted

---

[1]P.T. McGrath, *Newfoundland in 1911* (London, 1911).

upwards of seventy percent of all Newfoundland catches.[2] Geographically, the range of northern cod extends from Hopedale, Labrador (55° 20' N) southeast along the Northeast Newfoundland Shelf to include the northern half of the once biologically rich Grand Bank off southeastern Newfoundland (46° 00' N) (see figure 1).

Reported harvests appear to have been less than 100,000 tonnes until the early nineteenth century, increasing to as much as 300,000 tonnes in the 1880s and 1910s before declining to less than 150,000 tonnes in the mid-1940s.[3] Following the expansion of European-based factory trawlers in the late 1950s and early 1960s, particularly in the virtually unfished offshore waters off southeastern Labrador (e.g., Hamilton Bank), reported catches increased dramatically to a historical maximum of 810,000 tonnes in 1968 before collapsing in equally dramatic fashion to 1977 when Canada extended its fisheries jurisdiction to 200 miles. Controlled in part by the Total Allowable Catches (TACs) established by the Canadian government (the inshore component of the northern cod fishery had never been limited by catch quotas), reported catches increased gradually to a post-1977 high of 268,000 tonnes in 1988 prior to the imposition of a moratorium on the northern cod fishery in July 1992.[4] In early 2002, northern cod have shown no signs of recovery, the population having been estimated to be less than 0.5% of its size in the early 1960s.[5]

Spatial changes in the exploitation of cod appear to have been largely the result of discovery, war, territorial expansion, increased occupation of sites suitable for drying fish and declining catch rates. During the first eight decades of discovery in the 1500s, France, Spain, Portugal and England made increased use of the bays along the east coast of Newfoundland between Cape Bonavista and Cape Race. France also gradually increased its shore fishery to include the *Petit*

---

[2]J.A. Hutchings and R.A. Myers, "The Biological Collapse of Atlantic Cod off Newfoundland and Labrador: An Exploration of Historical Changes in Exploitation, Harvesting Technology, and Management," in Ragnar Arnason and Lawrence Felt (eds.), *The North Atlantic Fisheries: Successes, Failures and Challenges* (Charlottetown, 1995), 37-94.

[3]*Ibid.*

[4]J.A. Hutchings and R.A. Myers, "What Can Be Learned from the Collapse of A Renewable Resource? Atlantic Cod, *Gadus morhua*, of Newfoundland and Labrador," *Canadian Journal of Fisheries and Aquatic Sciences*, LI (1994), 2126-2146.

[5]G.R. Lilly, *et al.*, "An Assessment of the Cod Stock in NAFO Divisions 2J+3KL," Department of Fisheries and Oceans, Canadian Stock Assessment Secretariat Research Document 2001/044 (2001).

Nord from Cape Bonavista to Quirpon and then southward along the Newfound-land and Labrador shores of the Strait of Belle Isle.[6] From sixteenth-century beginnings, France continued to expand its fishery along the south coast in the early seventeenth century. The bank fisheries of the sixteenth and seventeenth centuries yielded relatively small catches compared with the shore fishery and were prosecuted primarily on the southern Grand Banks as well as on the smaller banks off the southeast coast (e.g., St. Pierre Bank, Green Bank and Whale Bank).[7]

The largest spatial expansion of the Newfoundland fishery for Atlantic cod began in the early nineteenth century with a dramatic increase in fishing effort northward along coastal Labrador. According to Gosling, reported catches of cod from Labrador increased from 24,750 quintals (1 quintal = 112 lbs.) in 1806 to 134,580 quintals, taken from between Cape Charles and Sandwich Bay, in 1820. Expansion of the Labrador fishery began in earnest in the 1860s. Individuals testifying to a Committee of the Newfoundland House of Assembly in 1856 reported that 700 Newfoundland vessels sailed to Labrador each season, fishing between Blanc Sablon and Cape Harrison, the northern limit of the Labrador fishery until the late 1850s.[8] Through the 1860s, schooners ventured north of Cape Harrison to Hopedale, and by 1870 over 500 Newfoundland vessels are reported to have passed north of Hopedale, with 145 being counted on a single day. By the mid-1870s, the fishery had expanded north of Cape Mugford. By the 1890s, some schooners were venturing north of Ramah to the northern tip of Labrador. Our objective here is to explore possible reasons for this expansion.

The Economic Environment of the Early Nineteenth Century

The historiography portrays the Napoleonic era as a watershed in the development of Newfoundland society. Before the 1790s, the seasonal nature of the migratory industry required only a handful of residents to look after fishing equipment and property, and people found little in the way of other resources to support much out-of-season residence. The Napoleonic wars, however, disrupted the migratory

---

[6]R. de Loture, *Histoire de la grande pêche de Terre-Neuve* (Paris, 1949; English trans., Washington, DC, 1957).

[7]C. de La Morandière, *Histoire de la pêche française de la morue dans l'Amérique septentrionale* (Paris, 1962); and J.H. de La Villemarqué, *La pêche française de la morue du XVIe au XVIIIe siècle dans l'Atlantique du nord-ouest* (Paris, 1991).

[8]W.G. Gosling, *Labrador: Its Discovery, Exploration and Development* (Toronto, 1910).

fishery, gave British settlers in Newfoundland a near-complete monopoly in the world market for saltfish and encouraged rapid permanent settlement by immigration between about 1810 and 1815. The postwar return of more competitive markets ended this unusual pattern of immigration. The Newfoundland population developed slowly over the generations, constantly searching for additional resources to allow permanent residence. The expansion of domestic marine industries, especially the seal hunt and the Labrador fishery, solved the initial post-1815 crisis. The literature suggests that the Labrador fishery, although never very profitable on its own, was fortuitous because it solved the problem of unused capacity in the schooner fleet employed in the spring seal fishery and provided complementary earnings for people unable to support themselves on the Newfoundland inshore fishery alone.[9]

---

[9]The seminal work is Shannon Ryan, "Fishery to Colony: A Newfoundland Watershed, 1793-1815," in P.A. Buckner and David Frank (eds.), *The Acadiensis Reader, Volume One: Atlantic Canada Before Confederation* (Fredericton, 1985), 130-148. Ryan also produced the most careful examination of the place of the Labrador fishery in the more general economic history of the Newfoundland cod fisheries. See Ryan, *Fish Out of Water: The Newfoundland Saltfish Trade, 1814-1914* (St. John's, 1986), 46-55; and his *The Ice Hunters: A History of Newfoundland Sealing to 1914* (St. John's, 1994), *passim*. Ryan's interpretation influenced heavily Sean T. Cadigan, *Hope and Deception in Conception Bay: Merchant-Settler Relations in Newfoundland, 1785-1855* (Toronto, 1995). Other works mention the incidence of the Labrador floater fishery without offering much explanation about why it should have developed in the first place. These studies suggest that English exploitation of Labrador coastal waters had been a part of the English migratory fishery from the early eighteenth century. American and French treaty rights and encroachments in the waters around Newfoundland pushed later colonial fishers from there onto the Labrador coast. The best explanation that can be extracted from this work is the economic one of solving the problem of unused capacity for sealing schooners. See B. Wade Colbourne and Robert H. Cuff, "Labrador," and Cuff, "Labrador Fishery," *Encyclopedia of Newfoundland and Labrador, Volume 3* (St. John's, 1991), 203-216 and 225-230; Cuff, "Floaters," *Encyclopedia of Newfoundland and Labrador, Volume 2* (St. John's, 1984), 220-221. W.A. Black's study of the economic geography of the floater fishery suggests that it developed largely by imperial encouragement as a means to ensure British claims to the coast between 1765 and 1792. Black made the intriguing claim that poor fisheries on the Grand Banks forced some banking schooners onto the Labrador coast after 1815. But Black did not identify whether or not these were Newfoundland or British migratory vessels. Furthermore, Black's argument would not explain the growth of French and American effort on the same banks at that time. See Black, "The Labrador Floater Codfishery," *Annals of the Association of American Geographers*, L, No. 3 (1960), 267-268. Black drew his explanation of imperial strategy from Gosling, *Labrador*, 379-414. Prowse offers similar arguments. His account implies that the constant northward development of the Labrador cod fishery grew out of the search for new fishing grounds. Prowse did not indicate whether or not this search was the result of something being wrong with older fishing grounds. See D.W. Prowse, *A History of Newfoundland from the English, Colonial*

Figure 1
Map of Newfoundland and Labrador

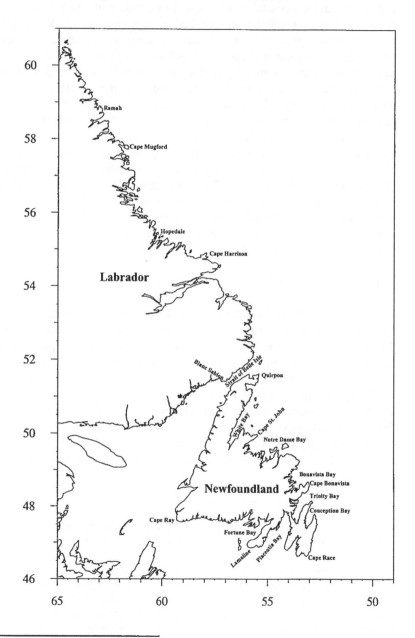

and Foreign Records (London, 1895), 586-607.

Local popular histories of communities in Conception and Bonavista bays suggest that growing cod scarcities in inshore waters, relative to increasing human populations in local coastal areas, led to the expansion of effort in the Labrador fishery.[10] A more explicit ecological historical perspective on the relationship between humans and northern cod may not allow us to continue to accept as historically positive the development of the Labrador fishery.[11] That fishery might rather have reflected a growing ecological crisis in the inshore Newfoundland fishery that began with the increase of the resident human population by 1815. At best, the need to expand into Labrador waters suggests that there were easily-reached limits to the carrying capacity of the island's cold-ocean coastal ecosystems for human populations within a social and economic infrastructure dominated by merchant capitalism. At worst, the attendant increase in fishing effort in inshore waters may have reduced the abundance of local cod stocks. Reductions in local cod stocks may have intensified the growing imbalance between human and cod populations caused by the former's demographic increase in numbers. The spatial shift to the Labrador fishery may have been as much a response to such an imbalance in the inshore fishery as it was to the need to find a use for sealing schooners during the summer.

The Crisis of Inshore Cod in the Nineteenth Century

There is some evidence that fishers may have been exhausting discrete bay or sub-stocks of northern cod in inshore waters in the first half of the nineteenth century. The diaries of William Kelson, the agent of the fish-merchant firm of Slade and Kelson, show that its employees and clients noticed a decline in the availability of cod in Trinity Bay in these years. Slade and Kelson's clients found so little cod for trading that the firm began to restrict credit almost yearly throughout the 1840s and 1850s. Newspaper accounts support the view that the inshore fishery was failing regularly in Conception, Trinity and Bonavista bays in the first half of the nineteenth century. Reports from Newfoundland newspapers for the period 1856 to 1880 further suggest that failure continued to plague fishers who exploited the

---

[10]Eric Martin Gosse, *The Settling of Spaniard's Bay* (St. John's, 1988), 42-43; and John Feltham, *Northeast from Baccalieu* (St. John's, 1990), 69 and *passim*.

[11]The related but distinct fields of ecological and environmental history are still fairly undeveloped in Canadian historiography with the notable exception of First Nations' history. Samples from some of the existing work, including Farley Mowat's *Sea of Slaughter*, and a good national and international bibliography can be found in Chad Gaffield and Pam Gaffield (eds.), *Consuming Canada: Readings in Environmental History* (Toronto, 1995). A model ecological history is Arthur F. McEvoy, *The Fishermen's Problem: Ecology and Law in the California Fisheries, 1850-1980* (Cambridge, 1996).

longest-used inshore fishing grounds along the Newfoundland coast. These reports deemed thirteen of those twenty-five years to have been general failures.[12]

It is reasonable to assume, then, that the conditions of scarcity noted by Kelson were an important reason for the fluctuations in Newfoundland's saltfish exports in these years. The clients of Kelson's firm might have been witnessing the effects of their over-exploitation of distinct inshore stocks of Atlantic cod in Newfoundland. Another way of stating this hypothesis is in the form of a question: How much of the inter-annual variation in inshore Newfoundland catches can be attributed to fishing mortality? Inshore catch levels have been, of course, also influenced by factors that affect the availability of cod to the fishing gear (e.g., water temperature, wind and abundance of prey) and by inter-annual changes in natural sources of mortality (e.g., environmental variation in recruitment and predation of cod by seals).

Overfishing in the present context refers to the over-exploitation of sub-components of the northern cod stock complex. Little historical data exist to allow a precise examination of these sub-components, but twentieth-century evidence is suggestive. Although northern cod are managed as a single unit, tagging experiments, genetic studies, spawning distributions and local knowledge of fishermen provide evidence of separate inshore and offshore stocks.[13] A study that

---

[12]Sean Cadigan, "The Moral Economy of the Commons: Ecology and Equity in the Newfoundland Cod Fishery, 1815-1855," *Labour/Le Travail*, XLIII (Spring 1999), 9-42; and Cadigan, "Failed Proposals for Fisheries Management and Conservation in Newfoundland, 1855-1880," in Dianne Newell and Rosemary Ommer (eds.), *Fishing People, Fishing Places: Issues in Canadian Small-scale Fisheries* (Toronto, 1999), 147-169. Cadigan examined every surviving issue of four St. John's newspapers: *The Public Ledger*, *The Newfoundlander*, *The Newfoundland Express*, and *The Courier*, and recorded any comments they made about general fishing conditions, locations of failed and successful catches, types of gears used and explanations offered. He supplemented this data by reading the only papers from outside St. John's to have published in the period for which copies exist: *The Standard* (Harbour Grace) and *The Twillingate Sun*.

[13]W. Templeman, "Comparison of Returns from Different Tags and Methods of Attachment Used in Cod Tagging in the Newfoundland Area, 1954 and 1955," *ICNAF Special Publications*, IV (1963), 272-287; W.H. Lear, "Discrimination of the Stock Complex of Atlantic Cod (*Gadus morhua*) off Southern Labrador and Eastern Newfoundland, as Inferred from Tagging Studies," *Journal of Northwest Atlantic Fisheries Science*, V (1984), 143-159; C.T. Taggart, *et al.*, "The 1954-1993 Newfoundland Cod-tagging Database: Statistical Summaries and Spatial-temporal Distributions," *Canadian Technical Report of Fisheries and Aquatic Sciences* (1995), 2042; D.E. Ruzzante, *et al.*, "Genetic Differentiation between Inshore and Offshore Atlantic Cod (*Gadus morhua*) off Newfoundland: Microsatellite DNA Variation and Antifreeze Level," *Canadian Journal of Fisheries and Aquatic Sciences*, LIII (1996), 634-645; J.A. Hutchings, R.A. Myers and G.R. Lilly, "Geographic Variation in the Spawning of Atlantic Cod, *Gadus morhua*, in the

estimated inshore exploitation rates from 1948 to 1992 identified four such stock sub-components: bay stocks, i.e., sub-populations that spawn and overwinter in the deep arms of bays such as Trinity and Bonavista Bays; headland stocks, i.e., sub-populations that overwinter in deep water off headlands such as Cape Bonavista; offshore migrants that overwinter on the edge of the continental shelf and migrate inshore to feed in late spring and early summer; and offshore residents that do not migrate inshore.[14]

Inshore catches of northern cod may be comprised of different proportions of bay, headland and offshore cod. Although the proportional representation of each stock sub-component in the catches is not known, it probably varies seasonally. Based upon interviews with fishermen from the Southern Shore north to Notre Dame Bay, bay cod appear to comprise the bulk of catches in early spring, late autumn, and winter, while offshore cod probably dominate catches in June, July, and August (during which bay and headland cod are also presumably caught).[15]

The inshore northern cod fishery can be characterized, then, as a mixed-stock fishery. Mixed-stock fisheries pose significant problems to maintaining sustainable, long-term harvests. If some components of the stock complex are more easily fished than others, they become relatively easy to over-exploit, or possibly eliminate.[16] Given their presumed low numbers, the bay and headland sub-components of the northern cod stock may have been vulnerable to over-exploitation at catch levels that would have had comparatively little effect on the offshore migrant sub-component. Mark-recapture studies have been used to quantify the fishing mortality experienced by cod thought to be part of one or

---

Northwest Atlantic," *Canadian Journal of Fisheries and Aquatic Sciences*, L (1993), 2457-2467; and B. Neis, *et al.*, "Fisheries Assessment: What Can Be Learned from Interviewing Resource Users?" *Canadian Journal of Fisheries and Aquatic Sciences*, LVI (1999), 1949-1963.

[14]R.A. Myers, N.J. Barrowman, and J.A. Hutchings, "Inshore Exploitation of Newfoundland Atlantic Cod (*Gadus morhua*) since 1948 as Estimated from Mark-recapture Data," *Canadian Journal of Fisheries and Aquatic Sciences*, LIV (1997), supplement 1, 224-235.

[15]B. Neis, *et al.*, "Northern Cod Stock Assessment: What Can Be Learned from Interviewing Resource Users?" Department of Fisheries and Oceans, Atlantic Fisheries Research Document 96/45 (1996); A.J. Potter, "Identification of Inshore Spawning Areas: Potential Marine Protected Areas?" (Master in Marine Management Project, Marine Affairs Programme, Dalhousie University, 1996); and unpublished data of J. Hutchings and M. Ferguson.

[16]C.W. Clark, *Mathematical Bioeconomics* (2nd ed., New York, 1990).

more bay stocks. They estimated that forty percent of the stock biomass of bay cod was removed annually by fishing in the late 1940s.[17] This is well in excess of the eighteen percent harvest rate considered to ensure long-term sustainability. Thus, rather than the over-exploitation of the northern cod stock, the result of which has been a commercial fishing moratorium,[18] it is the potential overfishing of bay and headland cod in the nineteenth century that we wish to address.

Figure 2
Newfoundland's Saltfish Exports in Hundredweights, 1805-1884

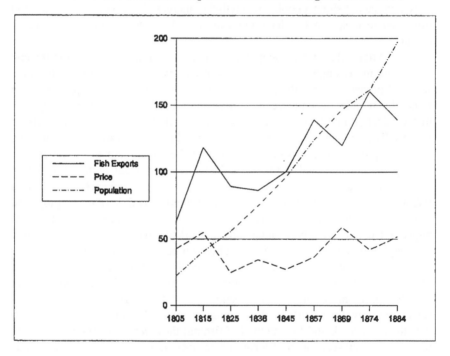

Note: (1 = 10,000), Fish Prices (US $10) and population ('000).

Sources: Shannon Ryan, *Fish Out of Water: The Newfoundland Saltfish Trade, 1814-1914* (St. John's, 1986), 250-260; and Daniel Vickers, Codfish Price Index for Philadelphia.

---

[17]Myers, Barrowman and Hutchings, "Inshore Exploitation."

[18]Hutchings and Myers, "What Can Be Learned."

The peaks in Newfoundland's total fish exports (figure 2) might be taken as a contradiction of the possibility that bay and headland cod were being overfished if considered alone. But fluctuations in Newfoundland salt cod exports, while the human population increased steadily throughout the nineteenth century, suggest a basic ecological problem. Although Newfoundland's overall total exports of salt cod tended to increase to 1884, there were dramatic drops between 1815 and the late 1840s, the 1860s, and after 1874. Prices for salt cod dropped in the post-1815 international depression, but the fluctuations in Newfoundland's total exports of salt cod do not appear to bear much relation to those in prices. Newfoundland fishing people, if anything, increased the volume of salt cod produced to make up for declining values in the first half of the nineteenth century.

After 1815, fishing people made up for declining incomes from the cod fishery by increasingly relying on the seal fishery. There are no reliable time series of prices to assess the relationship between exports of seal products and prices, but popular dependence on the seal hunt for earnings nevertheless suggests that a reduction in effort is unlikely to explain declining seal product exports up to 1884. The growing public debate about the need for conservation legislation due to over-exploitation of seal herds throughout the second half of the nineteenth century implies that seal herds were declining.[19] Constant human population growth meant an increase in effort in the fisheries even as exports fell. Besides fishing, sealing, and supplementary farming, there was little other employment available to Newfoundlanders from 1815 to 1884. The percentage of the total population employed in Newfoundland that worked directly in the fishery ranged from eighty-nine percent in 1857 to eighty-two percent in 1884 (figure 3).

A number of factors may have masked the negative impact of the growth of the resident fishing industry on inshore mixed cod stocks. One element was the introduction of more intensive harvesting gears. In the 1840s, the use of cod seines and trawl lines (called bultows at the time) had become more widespread than the older hook-and-line method of fishing; they were essential to the initial recovery and subsequent short-term maintenance of export levels. By the 1860s fishers were using gill nets. The introduction of newer gears tended to improve catches for a short period, but declines soon returned, prompting the use of even more intensive gears. The recovery of saltfish exports in the 1870s coincided, for example, with the introduction of cod traps and expansion into newer fishing grounds further offshore, especially in the bank fishery.[20]

---

[19]Ryan, *The Ice Hunters*, 98-100, 105-106 and 111-117.

[20]Cadigan, "Failed Proposals."

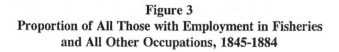

Figure 3
Proportion of All Those with Employment in Fisheries
and All Other Occupations, 1845-1884

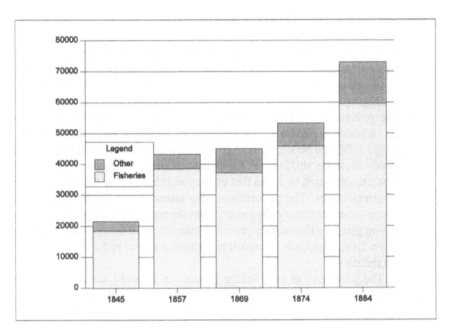

Source: Newfoundland, *Census of Newfoundland*, 1845-1884.

Although the 1815 peak in saltfish exports coincided with the end of the Napoleonic era, the export peaks of 1857 and 1874 were not associated with specific market-driven factors that would have stimulated catches. These peaks were, however, associated with a dramatic expansion of fishing effort along coastal Labrador, beginning with the six-fold increase in catches between 1814 and 1829. One estimate suggests that the total catches of cod from Labrador increased from 24,750 quintals in 1806 to 134,580 quintals, taken from between Cape Charles and Sandwich Bay, in 1820. The northward expansion of the Labrador fishery began in earnest in the 1860s. Individuals testifying to a Committee of the House of Assembly in 1856 reported that 700 Newfoundland vessels sailed to Labrador each season, fishing between Blanc Sablon and Cape Harrison, the northern limit of the Labrador fishery until the late 1850s. Through the 1860s, schooners ventured north of Cape Harrison to Hopedale, and by 1870 over 500 Newfoundland vessels are reported to have passed north of Hopedale with 145 being counted on a single day. By the mid-1870s, the fishery had expanded north of Cape Mugford. By the 1890s, some schooners were venturing north of Ramah to the northern tip of Labrador.[21]

The Expansion of the Labrador Fishery

A growing imbalance between Newfoundland inshore fishing people and the marine resources upon which they depended led to growing demands for public relief by people such as the clients of Slade and Kelson. The House of Assembly responded by trying to stimulate growth in other industries, especially in forestry. While the government wanted outport people to produce a wide range of wood products, it hoped in particular that shipbuilding would generate much employment. In the 1850s the Newfoundland government encouraged local shipbuilding by a subsidy of twenty shillings per ton to a maximum of fifty tons on any vessels constructed in shipyards or docks that employed at least seven able-bodied poor men at current wages. The government fully intended that shipbuilding would draw labour out of the fishery altogether.[22] But the encouragement to small-vessel construction given by the subsidy provided many inshore fishers with the means to find new fishing grounds to exploit in Labrador waters rather than withdraw from the industry.

The protection of new fishing grounds in coastal Labrador waters for Newfoundland fishers from French interlopers preoccupied colonial officials. The local government was especially vigilant against imperial concessions of further treaty rights to the French. The House of Assembly reacted angrily in 1857, for example, to British proposals to give the French freer access to Labrador waters.[23] Newfoundland merchants and government officials continued to blame inshore fishing failures on the French as part of a strategy to secure imperial support for proposed legislation which would deprive the latter of access to Newfoundland-caught herring bait. The hope in Newfoundland was that the loss of bait supplies would end the profitability of the French fisheries off Newfoundland's coasts, and therefore French treaty rights to the island's shores.[24] Advancing the notion that

---

[21]Gosling, *Labrador*, 413-425.

[22]*The Public Ledger* (St. John's), 8 and 11 April 1856; and *The Courier* (St. John's), 19 April 1856.

[23]*The Public Ledger* (St. John's), 3 March 1857.

[24]For an examination of the growing popularity in colonial politics of blaming the French and other "foreigners'" fisheries for declining inshore catches see Sean Cadigan, "A 'Chilling Neglect': The British Empire and Colonial Policy on the Newfoundland Bank Fishery, 1815-1855" (paper presented to the Canadian Historical Association annual

cod migrated inshore in a northerly direction from offshore banks, merchants such as C.F. Bennett blamed French bank fishing for the depletion of cod inshore. Over-exploitation of herring stocks on the south coast of Newfoundland by island fishers who made quick money by trading in bait was the further result of the French industry. While the reasons Bennett assigned for an ecological crisis in inshore Newfoundland waters might be questioned, there is no doubt that he perceived a problem. Although Newfoundland residents had on "their once famous fishing grounds, and almost within sight of their homes" been able "to catch...cod-fish in abundance," Bennett claimed, they now had "to proceed to the Labrador Coast, 500 to 1000 miles from their homes, for an inferior class of cod."[25]

The regulation of the herring trade and the protection of the Labrador cod fishery were not simply efforts by the Newfoundland government to undercut French competition in the cod fishery generally. Shortfalls in the inshore Newfoundland cod fishery, and the success of the south-coast bait trade, suggested the potential for the shift in some resident effort into a colonial herring fishery. An autumn commercial herring fishery had developed as an extension of the summer cod fishery. But as early as 1855 newspapers had been reporting that the Labrador herring catches were falling off. More concerned with keeping herring production up, the press advocated switching effort to the west and south coasts of the island although conceding the superior nature of autumn-caught Labrador herring to "our winter or spring fish."[26]

Colonial efforts to regulate the bait trade in herring emphasized to many in Newfoundland that the herring fishery had begun to absorb resident effort that might have been exceeding the capacity of the inshore cod fishery. The Newfoundland government's worries about foreign inroads on the Labrador cod fishery reflected the manner in which island residents had come to count on it to make up for declining inshore catches. A good Labrador catch in 1858 forestalled the disaster of a bad inshore result.[27] In Trinity Bay, for example, "the craft belonging to that place engaged in the Labrador fishery, have returned tolerably well fished and will make up a middling voyage. In Trinity the fishery has not been so good

---

meeting, Montréal, August 1995); and Cadigan, "A Shift in Economic Culture: The Impact of Enclave Industrialization on Newfoundland, 1855-1880" (paper presented to the Atlantic Canada Studies Conference, Moncton, NB, May 1996).

[25]C.F. Bennett to Colonial Secretary John Kent, 1858, reprinted in *The Newfoundlander* (St. John's), 14 January 1861.

[26]*The Newfoundlander* (St. John's), 24 December 1855.

[27]*Ibid.*, 16 August 1858.

as could be wished."[28] Reports on poor inshore fisheries credited both the Labrador cod and the herring fisheries for making up the shortfalls throughout the northeast coast, parts of Placentia Bay and the southern shore of the Avalon Peninsula.[29]

Newfoundland merchants, such as Stephen March, who invested in the Labrador fishery protested against government attempts to save money by ending a summer patrol service that policed against French encroachments. March pointed out that northeast-coast Newfoundlanders needed the waters off Labrador as a preserve to resort to when their own waters failed them.[30] Fishers who had boats large enough to do so left for the Labrador fishery as soon as it was clear that they would find little fish in inshore waters. Conception Bay residents dominated the rush for the Labrador coast and wanted a government steamer to police those waters to make sure that the French were not intruding.[31] The French were most likely to spill over into southern Labrador waters just beyond their treaty area on the French Shore. Conception Bay fishers found that catches around Triangle, Square Island and Dead Island had begun to fall off as early as 1860, so foreign competition was unwelcome. Otherwise, the Conception Bay effort had to move northward beyond Indian Tickle to find good fishing.[32]

Throughout the early 1860s the press continued to advocate the development of the herring fishery to supplement the cod fishery.[33] Diversification into herring production and expansion of the spatial range of exploitation by inshore cod fishers became the main strategies the press and government advocated as the best means to address failing cod catches from older inshore waters. The Conception Bay press suggested that inshore fishers abandon their small boats for jack boats – larger vessels that would increase their range.

---

[28]*The Courier* (St. John's), 9 October 1858.

[29]*The Newfoundlander* (St. John's), 16 August 1858.

[30]Provincial Archives of Newfoundland and Labrador (PANL), GN2/2, Incoming Correspondence of the Colonial Secretary, 1959, S. March to Acting Colonial Secretary Edward Shea, St. John's, 20 May 1859; see also a similar letter from Thomas R. Crockwell to Shea, St. John's, 18 May 1859. March had held the contract to provide a police boat for the Labrador coast in the past.

[31]*The Standard* (Harbour Grace), 6 June 1860; *The Newfoundland Express* (St. John's), 4 August 1860.

[32]*The Standard* (Harbour Grace), 12 September 1860.

[33]*The Newfoundlander* (St. John's), 5 May 1862; and *The Courier* (St. John's), 7 May 1862.

Although these jack boats have been associated with fisheries on the south coast of Newfoundland, contemporaries used the term interchangeably with that used for a similar boat employed on the northeast coast: the bully. Jacks and bullys were decked, two-masted, schooner-rigged boats that ranged from five to twenty tons burthen. While most studies suggest that Newfoundland fishers sailed to the Labrador coast in schooners of at least thirty tons, the larger jacks sailed there as well.[34]

It may be that the term "jacks" was used to describe vessels of around twenty tons that could actually make the voyage on their own from Newfoundland to Labrador. Much smaller ones, in the five-to-ten-ton range, might be the bullys described by Nicholas Smith as being used in the stationer fishery at Labrador. Such bullys would have allowed stationers to range further on headland fishing grounds, just as they did for inshore fishers in Newfoundland bays.[35] Effort increased so quickly in the Labrador herring fishery that the legislature passed an act in 1862 that prevented the use of seines to trap herring schools in coves. The new law banned the use of herring nets with meshes smaller than two and three-eighths inches.[36] The government directed this measure largely at American and Nova Scotian fishermen who used small-meshed nets to pen herring so that none had a chance to escape. Some Newfoundlanders fishing on the Labrador coast had become so upset with Nova Scotians that they had torn the latter's gear from the water and chased them off the coast. Although committed to the principles behind the law, the Newfoundland government had little money to spend on its effective enforcement.[37]

The Labrador herring fishery made up for poor cod catches on the coast through the early 1860s even as the inshore fishery on the island continued to fail. In 1861 the cod fishery failed in many areas, while in 1862 only the "resource

---

[34]*The Standard* (Harbour Grace), 24 and 26 August 1863; Cuff, "Labrador Fisheries;" Cuff and M.F. Harrington, "Shipbuilding," *Encyclopedia of Newfoundland and Labrador, Volume 5* (St. John's, 1994), 168; "Bully" and "Jack," in G.M. Story, W.J. Kirwin and J.D.A. Widdowson (eds.), *Dictionary of Newfoundland English* (Toronto, 1992), 73, 271.

[35]See Nicholas Smith, *Fifty-two Years at the Labrador Fishery* (London, 1936), 79-80.

[36]*The Newfoundlander* (St. John's), 24 and 31 March 1862.

[37]*The Public Ledger* (St. John's), 1 April 1862 and 3 March 1863; and *The Newfoundlander* (St. John's), 15 May 1862.

funnel" of the Strait of Belle Isle provided good fishing.[38] The cod fishery on the southern Labrador coast was otherwise meagre. Labrador fishers had better catches in 1863 because they began to use more intensive cod seines rather than relying so much on baited hand-lining. The news of improvement was especially welcomed in Conception Bay because "we want something to make up for the shore fishery, which has almost been a total failure in this neighbourhood the present season."[39] The success with cod seines on the Labrador coast became one of the most important arguments used by the opponents of conservation measures in the cod fishery. They resisted government consideration of limiting the use of more intensive harvesting gears by arguing that only such gears maintained export levels.[40]

By the mid-1860s it had become common practice for fishers who had larger jack boats or schooners to race for Labrador as soon as it was clear that they would find little cod along the Newfoundland coast. The press generally credited the Labrador catch with maintaining overall colonial exports against declining inshore catches. In Conception Bay the Labrador fishery was "the only one the greater portion of our people are interested in, the shore fishery being a total failure for some years past."[41] Some fishers had come to depend on the herring fishery so much that they asked the legislature to amend its herring legislation to permit Newfoundlanders to bar the fish. The weight of public opinion, however, encouraged the government not to loosen this conservation measure.[42] Dependence was no guarantee of success; the failure of the Labrador herring and cod fisheries in 1864 led the government, on the advice of the

---

[38]On the manner in which the narrow Strait of Belle Isle tended to concentrate marine resources in one area see Ralph Pastore, "The Collapse of the Beothic World," *Acadiensis*, XIX (Fall 1989), 61.

[39]*The Newfoundlander* (St. John's), 21 August 1862 and 30 July 1863; *The Newfoundland Express* (St. John's), 10 October 1861, 9 September and 7 October 1862, and 28 July and 25 August 1863; and quote from *The Standard* (Harbour Grace), 23 September 1863.

[40]*The Newfoundlander* (St. John's), 11 January 1864.

[41]*The Newfoundland Express* (St. John's), 19 July 1864; *The Newfoundlander* (St. John's), 29 August 1864; and quote from *The Standard* (Harbour Grace), 3 August 1864.

[42]*The Public Ledger* (St. John's), 1 March 1864; and *The Newfoundlander* (St. John's), 17 March 1864.

churches, to declare "a day of humiliation and prayer in consideration of the widespread suffering arising from the failure of our fisheries."[43]

The spatial expansion of Newfoundland fishing effort fuelled local support for Confederation in the 1860s. The pro-confederate press attacked the supporters of an independent Newfoundland colony who felt that the island's fisheries could maintain it alone. "When we hear of 'rich resources in our fisheries,'" wrote one editor, "one is forcibly reminded that of late years these 'rich' resources have but too deplorably failed for the support of the population employed in them."[44] MHA (Member of the House of Assembly, Newfoundland) William Whiteway contended that Newfoundland could never prosper if it had to depend on its inshore and Labrador fisheries alone. Canada, he thought, might support the development of a Newfoundland bank fishery that could compete effectively with the French. As early as 1865, Whiteway declared that the southern Labrador fishing grounds used by Newfoundlanders were becoming inadequate for their support, as were older island inshore waters. Whiteway's support for Confederation came on the heels of a particularly bad year for the fishery in 1864: the inshore cod fishery had failed, the Labrador catch had fallen off, and the markets for herring proved poor. The government tried to respond by spending $2000 in subsidies to encourage a local bank fishery, and $1000 to encourage a local mackerel industry.[45] These expenditures were token amounts that did not have the desired effect. MHAs such as Whiteway looked to a union of British North America for the requisite support. Even if Newfoundland did not expand onto the banks, Whiteway felt that Canadian support for industrial development of the fisheries was essential. Only the Canadian government would be able to support a more vigorous programme of large-vessel construction that would allow local fishers to search out new fishing grounds.[46]

The St. John's Chamber of Commerce opposed Confederation because it felt that Canada was unlikely to take an interest in marine-resource development or worry much about improving the markets for salt cod. The Chamber otherwise agreed with Whiteway that the older fishing grounds resorted to by Newfoundlanders could not support the growing numbers who exploited them. Expansion of the banks' cod fishery and the development of a mackerel fishery were the alternatives proposed by the Chamber. The merchant body additionally agreed with Whiteway that the inshore fishing industry needed restructuring.

---

[43]*The Newfoundland Express* (St. John's), 31 December 1864.

[44]*The Newfoundlander* (St. John's), 5 December 1864.

[45]*The Courier* (St. John's), 28 January and 22 April 1865.

[46]*The Newfoundlander* (St. John's), 2 February 1865.

Fishers needed to invest more capital in larger, more wide-ranging vessels. Poor catches, argued the Chamber, resulted from the small boats used by most inshore fishers. The "fishermen are thereby restricted to their own immediate locality, and cannot resort to distant points, where fish may be abundant, and a saving voyage be readily secured."[47]

Newfoundland fish merchants felt that the local government could better direct its attention to providing better steam communication with the Labrador coast than towards the idea of Confederation. The St. John's Chamber of Commerce asked the government to send a steamer to the coast in early August to "bring back intelligence both of the result of the Summer fishery, and the prospect for the Fall [cod] & Herring fisheries." Merchants and fishers wanted better access to information about the Labrador fishery so that they might respond more efficiently to inshore failures. Conception Bay merchants had to balance the greater credit requirements of the outfit of Labrador schooners and jack boats against the likelihood of their successful catches.[48]

The strategy of redirecting effort from inshore waters around the island to the Labrador coast spread quickly from Conception Bay, although it was no guarantee of a successful season. A Bonavista Bay correspondent, for example, wrote at the beginning of the 1866 fishing season that

> we are all in the doldrums here of late, as the sailors say, being between two winds, or rather two minds; that is, whether to embrace the first favourable time and proceed right off to Labrador, in search of the recreant cod, who, it seems, has been smitten with...a great desire of seeing foreign parts, or rather to remain at home...until he, his codship, pleases to pay us a visit."[49]

Newfoundlanders generally waited for news of a good Labrador fishery "with great anxiety, remembering how vastly the importance of that fishery is increased by the flight of so large a portion of our Northern population to the Labrador coast."[50]

---

[47]*The Courier* (St. John's), 5 August 1865.

[48]PANL, GN2/2, 1866-1867, T. Rendell [President of the Chamber of Commerce] to Colonial Secretary J. Bemister, St. John's, 9 July 1866; and *The Standard* (Harbour Grace), 5 July 1865 and 10 January 1866.

[49]*The Standard* (Harbour Grace), 27 June 1866.

[50]*The Newfoundlander* (St. John's), 9 July 1866.

Newfoundland's growing dependence on the Labrador fishery led some to question the Chamber of Commerce's opposition to Confederation. The inshore fishery failed again in 1866 and the Labrador catch, although good, did not prove sufficient to make up the inshore shortfall.[51] The social distress that followed provoked two reactions. The supporters of Confederation first reiterated that only Canada had the resources to help inshore fishers to get out of their "miserable cockleshells of punts which can only creep close along the shore when all is calm, and must fly in and abandon perhaps the best fishing at the first ripple of a breeze." The expansion of vessel size would allow fishers to search for new fishing grounds. Canada might also support diversification into the herring and mackerel fisheries.[52]

The second reaction reflected the manner in which British authorities had previously been so willing to bargain away rights to the Labrador fishery. "Colonus," for example, feared that local hostility to Confederation would make Canadians the next rivals for Newfoundland access to Labrador fish:

> The Bank fishery is forsaken by us. The catch of Shore fish is diminishing. Labrador is our main support. Our condition is even *now* precarious. But, if Canada backed by the Lower provinces claim Labrador which is connected with Canadian Territory – and if the said claim...be acceded to by the Imperial Government – and if, in order to punish our anti-union proclivities, prohibitory or even high duties be imposed by the Confederation on our Labrador fishing material, what would be our condition *then*?[53]

Support for Confederation arose partially from fears that a united British North America would take control of Labrador and from hopes that Canada might support the increasing capitalization of the Newfoundland fishing industry. Contemporary observers were not sure that the need for a larger, more mobile inshore fishing fleet indicated that inshore cod stocks' ability to survive was being threatened by fishing effort. One advocate of larger vessels argued, for example, that fishers might find that "fish may be scarce in the place they reside, and perhaps a few miles on either side of their settlement it may be plentiful, but they cannot avail of that, their boats not being sufficiently large to take them to those

---

[51]*The Public Ledger* (St. John's), 25 September 1866.

[52]*The Newfoundlander* (St. John's), 17 September 1866 and 4 June 1867.

[53]*The Standard* (Harbour Grace), 14 November 1866.

places."[54] Another commentator remarked, however, that no one could doubt that most inshore fishers had to go much farther to find fish and stay out on the water for much longer periods of time to catch the same amount of fish as they might have years ago. "Whatever the cause may be, the fact admits of no question, that the Codfish along our coasts have deserted their former haunts, and to be successful in catching them the fishermen must be so fitted out as to be enabled to follow them into deeper water."[55]

The "stereotypical failure" of the 1866 inshore fishery and the shortfall of the Labrador catch did not convince anti-Confederates, who eventually won their battle against confederation in 1869, to change their minds.[56] Anti-Confederates feared Canadian influence over future local resource development. They preferred instead that Newfoundlanders alone improve productivity in the Labrador fishery by the introduction of larger, more mobile fishing craft. The Labrador fishery to that point had largely been a stationers' effort along the southern coast. While the fishery there had been poor in 1866, those who ventured north towards Cape Harrison had better voyages.[57] Merchants' schooners carried fishing families and crews with their small boats to the Labrador coast from the Strait of Belle Isle to Cape Harrington. The schooner crews fished along the coast while waiting to ship the stationers' catches at the end of the fishing season. As early as 1840 poor cod catches on the coast below Cape Harrison made the efforts of such crews' fisheries more important as they began to search for more

---

[54]*The Courier* (St. John's), 14 July 1866.

[55]*The Newfoundland Express* (St. John's), 27 September 1866.

[56]The Newfoundland government had been exploring the possibility that it could secure more control over marine resources by union with Canada. Confederation proposals faltered because few in Newfoundland could see much Canadian interest in better marine management. To the contrary, provinces such as Nova Scotia that also depended on marine resources were turning landward to embrace the National Policy. In an effort to maintain support, Newfoundland confederates began to promise federal backing for landward diversification as well as for better fisheries management. This strategy ironically fell into the hands of opponents of confederation, who argued that if, as the confederates suggested, the colony had interior resources to be developed, why should they be subject to Canadian influence? See Cadigan "A Shift in Economic Culture." For arguments about the sectarian reasons underlying the Confederate loss in the general election of 1869, see James Hiller, "Confederation Defeated: The Newfoundland Election of 1869," in James Hiller and Peter Neary (eds.), *Newfoundland in the Nineteenth and Twentieth Centuries: Essays in Interpretation* (Toronto, 1980), 123-147.

[57]*The Newfoundland Express* (St. John's), 10 August 1866.

productive grounds northward to Cape Ailik, and later to Cape Mugford.[58] The vessels' passengers there went ashore and fished very much like they would have in inshore waters at home. The editor of one newspaper recommended that the stationers should become "floaters" like the schooner crews because

> it frequently happens...that the fishery becomes unproductive there, but the people have no opportunity of coasting from place to place where the fish is to be found. We would strongly recommend that this fishery should be pursued in schooners of from fifty to seventy tons which would be far more easily worked than large vessels and could therefore visit all parts of the coast with the people.[59]

In addition to the development of the floater cod fishery, merchants continued to attempt to develop markets for herring, especially in Ireland. The poor quality of the cure frustrated such efforts, and the press began to demand the appointment of inspectors.[60] Conception Bay merchants and planters asked that the government put two steamers on a regular run to the Labrador coast so that they could better supply both the herring and cod fisheries there. Their own vessels could not afford to carry enough empty barrels to store herring properly. Although the pelagic herring was an unpredictable fish to catch compared to groundfish like cod, it was "of increasing importance, particularly as the cod fishery in many places at the Labrador appears to be subject to much uncertainty."[61]

During the winter of 1866 Conception Bay merchants built or purchased larger vessels so that more of their stationer-clients could become floaters at Labrador.[62] The floater strategy reflected a belief popular with the Chamber of Commerce that there was no problem with the erosion of fish stocks due to fishing but rather that fish were capricious migrants. If inshore fishers found that they could no longer catch much fish on long-accustomed grounds, then that must be

---

[58]P.W. Browne, *Where the Fishers Go: The Story of Labrador* (New York, 1909), 60. Although the floaters did not commonly go to Cape Mugford until the late 1880s, or beyond until the 1890s, there is no way to determine exactly when the first explorers pushed north. For a good account of one schooner master's early experiment with a floater fishery as far north as Ramah in 1877, see Smith, *Fifty-two Years*, 29.

[59]*The Standard* (Harbour Grace), 2 January 1867.

[60]*The Newfoundlander* (St. John's), 10 January 1867.

[61]*The Standard* (Harbour Grace), 20 March 1867.

[62]*Ibid.*, 5 June 1867.

because the cod had changed their habits. Larger boats would allow fishers to seek out the fish in their new habitat.[63] Taking heart from the manner in which the introduction of steam-driven trawler smacks led to catch recoveries by opening up new fishing areas off Great Britain, Newfoundland commentators argued that the introduction of larger vessels in the colony's industry would restore faith in "the inexhaustible harvest of the sea spread around our coasts."[64]

Throughout the late 1860s the press began to attribute poor inshore fisheries to laziness or the conservatism of small-boat fishers for not investing in larger, more mobile vessels that would allow them to range out from headland areas to the Labrador coast. The press never asked whether or not the problem was that such fishers did not have access to the capital such investment would require. More important, the press did not consider whether or not small-scale fishers' relative impoverishment was related to inshore cod scarcities, although there had been plenty of past commentary about how reluctant merchants were to extend credit to inshore fishing families for exactly such reasons.[65]

The Newfoundland government was more suspicious than the press that there were growing problems with scarcity on older fishing grounds. In 1867, as a result of pressure from the Chamber of Commerce, it commissioned HMS *Gannet* to survey that state of fish resources on the Labrador coast. The officers found that the average catch by Newfoundlanders was more than double in the northern fishery above Cape Harrington that of the southern effort below it. "Newfoundlander," a correspondent of a Harbour Grace newspaper, argued that

> it is indeed a natural consequence that the fish being pursued year by year, particularly by those enormous seines, and pursued too by a very large body of fishermen, should quietly forsake their old places and seek for new abodes in the vast waters, where they can enjoy peace and safety. This history at all events of our Northern Cod fishery shows this one grave fact, the Northern progress of the fish.[66]

---

[63]*Ibid.*, 28 August 1867.

[64]*The Public Ledger* (St. John's), 23 July 1867.

[65]*Ibid.*, 2 August 1867; *The Express* (St. John's), 20 and 29 August 1867 and 8 September 1868; *The Newfoundlander* (St. John's), 16 August 1867 and 7 August 1868; and *The Standard* (Harbour Grace), 17 July 1867.

[66]*The Standard* (Harbour Grace), 2 October 1867; and Gosling, *Labrador,* 413-414.

Newfoundland's Colonial Secretary, John Bemister, found little credibility in the notion that cod scarcities in southern waters were the result of the fish moving north to escape fishers. "The fact is," he argued in the House of Assembly, "the species is declining, though we don't choose to admit it. He could not tell what was the true cause, but the fact is that the fishery on the Southern part of the Labrador coast is getting short."[67] Despite the Colonial Secretary's conviction, the captain of HMS *Sphinx*, on patrol along the Labrador coast in 1868, reported that catches appeared to be down for other reasons than that the cod were in trouble. Fishers told him that new fishing on a bank to the south was preventing the migration of fish up through the Gulf of St. Lawrence.[68]

The floaters at Labrador felt that they could find better fishing grounds north along the Labrador coast. In 1869 and 1870 their intuition proved false, and the only good cod fishing that year proved to be in the Strait of Belle Isle. Some in the press argued for more technological investment in the fishery. This time, however, a telegraph system should be put into place so that fishers in any harbour on the coast could learn immediately where good and bad fishing conditions prevailed. Such a telegraph system would compensate for the poor steamer service operated by government.[69] The Chamber of Commerce reported that adverse ice conditions on the "Northern extremity of the Coast of Labrador" had frustrated the floaters, but hoped that this would not discourage in future "our most energetic and persevering fishermen resorting to that comparatively distant fishing ground."[70]

The Labrador fishery appeared to improve in 1871 as a result of the use of more seines. but the press still reported the final catch as less than average. Through 1873, reports from the Labrador fishery suggested that only those who could use more capital- and labour-intensive gears aboard larger vessels in a floater fishery enjoyed much success.[71] The success of the floaters suggested:

---

[67]*The Express* (St. John's), 31 March 1868.

[68]Newfoundland, *Journal of the House of Assembly of Newfoundland*, IV (1869), appendix, 509-544, "Report of Captain Parish, on the 1st. cruise [sic] of HMS *Sphinx*, between 5th June and 22 August, 1868."

[69]*The Express* (St. John's), 12 October 1869; and *The Newfoundlander* (St. John's), 26 August and 16 September 1870.

[70]*The Courier* (St. John's), 10 August 1870; and *The Public Ledger* (St. John's), 20 September 1870.

[71]*The Express* (St. John's), 15 July 1871 and 13 August 1872; *The Courier* (St. John's), 27 July 1872; and *The Newfoundlander* (St. John's), 1 August 1871, 23 August 1872 and 19 August 1873.

the necessity of abandoning as a sole reliance the fishing as pursued by those who fixed themselves at definite stations. We find as a rule that vessels that visit the Coast with what is termed "a roving commission" generally secure a fair voyage, and it is evident that this result can only be assured with any degree of certainty by the means being at hand to enable the fishermen to abandon unproductive grounds and resort to those places where better fishing is reported.[72]

More rapid communications with the Labrador coast by steam vessels allowed merchants and fishers to redirect floaters' effort more quickly despite complaints about the service. If the inshore fishery proved very bad, steamers might bring news of whether or not fishers could hope to make up the shortfalls by a late trip to the Labrador coast. For fishers already on the coast, the steamer could bring news of exactly where most fish appeared to be caught. In 1874 the House of Assembly appointed a select committee to consider means by which the government might hire faster steamers because "the migratory habits of fish being well known, it is, of course, matter of the utmost importance to all interested in the Labrador fisheries, that the fishermen on that coast should have prompt information as to where they may find them most plentiful."[73] The government's throne speech in February 1875 indicated satisfaction that its Labrador steamer allowed effort to concentrate on the most productive fishing grounds in 1874. The government was so pleased that it began to consider whether or not it could modify its shipbuilding bounties to encourage the building of vessels better suited to more rapid redeployment of effort based on better information provided by the steamers.[74]

The captain of the government steamer *Gulnare* reported at the end of the 1875 fishing season that only the floaters would have good voyages in the Labrador fishery.[75] The press clamoured for other small-boat fishermen to join the floaters by investing in larger vessels so that "they may seek their future around the coast or at Labrador."[76] By 1876 fishers from even the most recent areas of

---

[72]*The Newfoundlander* (St. John's), 2 September 1873.

[73]*The Express* (St. John's), 26 March 1874.

[74]*Ibid.*, 6 February 1875.

[75]*The Newfoundlander* (St. John's), 21 September 1875.

[76]*Ibid.*, 27 June 1876.

settlement on the northeast coast of Newfoundland, such as Green Bay, "who could get off for Labrador had left for there" during the fishing season.[77]

Floaters in the Labrador fishery continued to move northward in search of more productive fishing grounds in the 1870s. Their efforts led Professor Hind, who had been hired by Nova Scotian investor Mr. Ellershausen to survey mineral resources along the northern Newfoundland and Labrador coasts, to report to the colonial government that he had allegedly

> made the discovery that from Aillik [*sic* Ailik] Bay, about forty miles north of Cape Harrison and about one hundred and forty north of the Straits of Belle Isle, and in front of the islands which fringe the coast line for about three hundred miles, extending onwards towards Nain, there exists an immense range of fishing banks, frequented as feeding grounds by vast shoals of fish. These banks are situated at the distance of only about fifteen miles from the islands, and these will now probably attract the attention of our fishing craft, whose voyages have hitherto been confined to the mainland and islands. These banks give promise of becoming in a few years the future fishing grounds of Newfoundland. We are informed that over four hundred craft have passed Cape Harrison this year, but their operations have been entirely confined to the vicinity of the islands.[78]

Press reports added that these new fishing banks were about twenty to twenty-five miles off the Labrador coast and yielded cod that were much larger than those being caught in waters closer to shore. Further government surveys, argued the newspapers, should confirm Hind's report because

> The inquiry into this subject could not have been more season-able than now, when the numbers engaged in the Labrador fishery are increasing from year to year. We want more room to meet the enlarged demands on the fishing grounds. This fishery is becoming, or rather has already become, the mainstay of our trade; for it would be fearful to contemplate the position we must occupy at the present moment had the disastrous issue of the shore fishery last season found any parallel at Labrador. The conservation and enlargement of our resources in this

---

[77]*Ibid.*, 18 July 1876.

[78]*The Courier* (St. John's), 18 November 1876.

quarter therefore attain leading importance and cannot be too earnestly studied.[79]

Hind's report impressed the government, which hired him to prepare further studies. The governor's address to the opening of the next session of the legislature reported that the manner in which the Labrador fishery succeeded to counterbalance yet another poor inshore fishery supported the professor. "The best success was reaped on the Northern Labrador fishing grounds, and by floating crews," reported the governor. These facts "appear to indicate the advantage to be gained by Labrador men from directing their course further Northward than has heretofore been the practice, and from an increase in the number of vessels capable of cruising over the extensive line of coast."[80]

Hind divided the Labrador coast into sections according to how long parts had been visited by fishers from Newfoundland. For 120 years or so people had fished the coast as far north as Sandwich Bay. The southern coast from Sandwich Bay to Cape Harrison had been exploited by them for "a generation or more." Only in the fifteen years prior to 1876 had fishers gone north of Ailik, which was itself forty miles north of Cape Harrison. From Cape Harrison to Cape Mugford, Hind argued, there existed "an immense cod-fishing ground" larger than all other areas combined fished by Newfoundlanders and the French. Calling these grounds the Northern Labrador Fishing Grounds, Hind estimated their surface area at 5200 square miles. The large amount of icebergs in these waters was supposed to cause an abundance of small crustaceans on which the cod fed. Such plentiful prey, in turn, would guarantee the continued surfeit of cod. Hind disagreed with popular views that the cod migrated north from off the island of Newfoundland or that they were changing their migration patterns to avoid fishing effort. He felt that individual sub-stocks of cod seasonally migrated from fixed "deep water winter feeding grounds" to "the nearest coast spawning grounds." Hind reported that about 400 vessels from eighteen to ninety tons burthen had passed Cape Harrison in the past season. Almost all of the fishers aboard these vessels relied on jiggers, which were only useful in shallow water. Hind argued that the bountiful waters

---

[79]*The Newfoundlander* (St. John's), 22 December 1876.

[80]About 550 fishers from Bay Roberts petitioned the House of Assembly not to have anything more to do with Hind. These petitioners did not dispute Hind's opinions about better fishing grounds on banks off the northern Labrador coast. They rather felt that Hind was a charlatan who could not have visited all the parts of the coast in the short period he claimed. It was not so much that the petitioners took issue with the contents of the report, but rather they felt that he had not come by the information directly. The Bay Roberts people argued that their own practical experience of pushing the Labrador fishery northward could give the government the same information without "fitting out scientific expeditions." See *The Courier* (St. John's), 3 February and 24 March 1877.

of the Northern Labrador Fishing Grounds consisted of deeper water favoured by the biggest cod. Only by building larger vessels and adopting new fishing gear could the bigger fish be caught on those grounds.[81]

There was no headlong rush into large vessel construction, as might have been suggested by the Hind report. The herring fishery of the southern Labrador coast partially counterbalanced the problems of growing failures in the cod fishery there by the late 1870s. In 1877, for example, the press expressed its hopes that rumours of abundant herring on the coast were true. Inshore cod catches around the island had proved

> fearfully deficient...and we can hardly hope for more than that
> this may be mitigated by better work at Labrador than we have
> yet heard of from that quarter. A good take of herrings...would
> make appreciably against the ill-success that has thus far
> characterised operations in the Labrador Cod fishery.[82]

The final result of the Labrador cod fishery alleviated fears that it would join the inshore fishery in failure. The Chamber of Commerce reported that the southern Labrador fishery from Belle Isle to Round-hill Island had been very poor. But "the migratory vessels engaged on that coast were particularly fortunate, being able to follow the fish far North where large fares were secured. The stationary crews on the Northern portion of the coast reaped an abundant harvest" as well.[83]

Expansion of effort by space and gear buoyed the northern Labrador cod fishery. By 1877 the northern coast had become home each summer to stationers who, with even more northerly ranging floaters, enjoyed a much more successful fishery than those along the southern coast. But even the northern fishers met success only by relying on cod seines. By 1879 and 1880 northern fishers were setting cod traps as well as using seines to obtain good trips from Ailik to Hopedale.[84]

The expansion of effort into the herring and Labrador fisheries masked declines in the productivity of inshore waters. Such masking could not, however, hide the growing imbalance between people and marine resources. "The products

---

[81]Newfoundland, *Journal of the House of Assembly of Newfoundland*, 12, III (1877), appendix, 730-743, "Notes on the Northern Labrador Fishing Grounds, by H.Y. Hind, St. John's, 8th November, 1876."

[82]*The Newfoundlander* (St. John's), 18 September 1877.

[83]*The Public Ledger* (St. John's), 31 August 1877.

[84]*Ibid.*, 20 July 1877; *The Courier* (St. John's), 1 September 1877; and *The Newfoundlander* (St. John's), 29 July 1879 and 10 August 1880.

of the fisheries do not vary much in quantity," claimed one local newspaper, "but the inhabitants have increased considerably." The solution, according to this paper, was that "everything ought to be done to render the fisheries more productive" while government encouraged landward diversification.[85]

As Newfoundland fishers discovered newer, more northerly fishing grounds off the Labrador coast, their productivity continued to obscure the possibility that the need to find such grounds represented successive over-exploitation of older fishing areas. William Whiteway, for example, did not associate better overall catches in 1878 with industrial expansion by area. He rather attributed such catches to the bounty of

> an ocean fed by the great arctic current ever rolling and teeming with life...the dominant fish being the codfish, termed by naturalists when speaking of economics, as the fish of fishes, the prince of fishes. As the demand for this fish increased so did the supply, subject to those incidental fluctuations which ever occur in all nature's productions, whether upon the land or in the sea.[86]

Not everyone had been so confident as Whiteway that marine resources were inexhaustible. There had been considerable popular pressure on the Newfoundland government to take a more precautionary approach by adopting conservation measures rather than industrial expansion in response to failing inshore cod fisheries throughout the nineteenth century. In 1879, for example, the Society of United Fishermen asked the Newfoundland government to create a Department of Fisheries or a fisheries science commission to take charge of any further investigation of, and recommendations for, the fisheries. There was by that time, however, little interest in fisheries management. The press and government turned to landward diversification though a comprehensive government plan to open up mining and forest resources by a trans-island railway scheme in consequence of Newfoundland's long-standing failure to secure adequate imperial support for its fisheries initiatives and because of the development mania generated by the debates about confederation. Government fisheries policy otherwise concentrated on industrial expansion by encouragement of the bank

---

[85]*The Courier* (St. John's), 12 January 1878.

[86]*The Public Ledger* (St. John's), 4 May 1878.

fishery and the discovery of new fishing grounds off the northern coast of Labrador.[87]

The introduction of new harvesting technology and larger fishing craft, as well as the spatial expansion of the cod fishery northward, were the primary means of industrial expansion, especially on the northeast coast, through the 1880s. The use of cod seines, and increasingly newer cod traps, buttressed inshore catches. Although the new gears might improve inshore fishers' overall catches, they often yielded only large amounts of small fish that did not cure well. The result was that even when such small cod was available inshore, "the greater part of the boats...sailed for the Labrador" from places such as Trinity. On arrival at Labrador, these floaters would encounter growing numbers of stationers on the southern part of the coast below Cape Harrington. Many of the floaters would turn north from these more crowded fishing grounds. Floaters from Twillingate and Fogo habitually sailed north beyond the range of the government steamers so that no one knew about how they did until their return. In 1880 Twillingate floaters got some of the best catches in Labrador waters because ice had drifted into more southern coastal waters.[88]

The success of such floaters led to demands that the government reconsider changes that it had made to its shipbuilding subsidies. Intent on encouraging the construction of larger vessels suited to the offshore bank fishery, as well as the more northerly Labrador fishery, the colony had decided to give bounties only to vessels that were thirty tons burthen or more in 1876.[89] Most northeast-coast floaters required smaller vessels in the fifteen-to-twenty-ton range. The editor of Twillingate's newspaper argued that any fishers who wanted to build vessels that would allow them to fish beyond the unproductive inshore waters should be encouraged by government. The floaters faced great risks in the gales

---

[87]Cadigan, "Failed Proposals;" Cadigan, "A 'Chilling Neglect';" Cadigan, "A Shift in Economic Culture." Although the Whiteway government paid lip service to the need for a commission of inquiry on the fisheries, it took no action until 1888 when it established a Fisheries Commission. The 1888 Fisheries Commission, supervised by Adolph Neilsen, demonstrated greater interest in cod-stock enhancement by the artificial propagation of cod rather than conservation by restraining their depletion in the first place. See K.W. Hewitt, "The Newfoundland Fishery and State Intervention in the Nineteenth Century: The Fisheries Commission, 1888-1893." *Newfoundland Studies*, IX (1993), 58-80.

[88]*The Twillingate Sun*, 22 July, 26 August and 23 September 1880.

[89]Cuff and Harrington, "Shipbuilding," 170.

and sea ice that were common in northern waters; any help from government would only partially balance against such hazards.[90]

Flexibility was the most desirable feature of the smaller floater vessels. The press suggested that island fishers should no longer even try their luck in inshore waters at the beginning of the season. They should rather head directly for the southern Labrador coast and begin fishing their way northward. Encounters with ice often made the voyages of those who sailed directly north before beginning to fish unsuccessful. If every floater began to fish in the south first, they could always head north later in the season. But such a strategy was impossible if fishers' wasted time initially fishing inshore around Newfoundland.[91]

Although they might begin in southern waters, larger numbers of floaters found that the search northwards was becoming routine. Every year brought more entrants because of the "scarcity of fish in the waters nearer home."[92] No longer was it the case that floaters could depend on southern waters: "within the past few years it has been found necessary to go hundreds of miles further north than formerly in order to secure a catch."[93] Through the early 1880s the floater fishery usually reported the best catches and commonly remarked that stationers did not fare so well on the Labrador coast. The press continued to pressure the government to amend its shipbuilding legislation to assist northeast-coast fishers who wished to become floaters by building more affordable jack-sized vessels. The government responded in 1883 to allow bounties to floater craft that were under thirty tons.[94]

The floaters' voyages north were not always the most successful. In 1886, for example, stationers from Grady to Battle Harbour had a good year, while the fishing proved better from there to Cape Harrison, and best from Turnovic to Nain. The floaters who went further north "returned from down the shore with little or no fish, and will be compelled to come home after the Summer without anything to meet the expenses." To make matters worse, the herring fishery had failed them as well.[95] Nonetheless, the northward extension of fishing

---

[90]*The Twillingate Sun*, 30 June and 21 and 28 July 1880.

[91]*Ibid.*, 25 August 1881.

[92]*Ibid.*, 7 July 1882.

[93]*Ibid.*, 21 October 1882.

[94]*Ibid.*, 7 September and 3 November 1883.

[95]*Ibid.*, 21 August and 4 and 11 September 1886. Two problems characterized the herring fishery. Poor packing had always hurt the reputation of the Labrador product in fish markets (Browne, *Where the Fishers Go*, 58-59). Labrador fishers also found that

effort continued. In 1887 floaters found little fish in southern Labrador coastal waters. Craft from Twillingate went so far north that year that the town's newspaper remarked that

> Of late years craft go further North than was ever before known, many of them going over a hundred miles north of Cape Harrington, which is several hundreds of miles from here. Places that were formerly known as prime berths for catching loads of fish further southward are now almost deserted by the "finny tribe" which appear to be seeking more secluded places for restoration, and nearly every year we hear of vessels going further and further north where fish seem to be quite plentiful.[96]

Failure, however, met these northern fishers again in 1888, as the cod appeared in southern waters but not in the north.[97]

The steady expansion of the Labrador fishery masked declining catches and catch rates (a metric of fish abundance) even as it proved to be a response to such downturns.[98] To explore this hypothesis further, we used information from the Newfoundland censuses between 1845 and 1911 to quantify harvesting capacity, fishing effort, catch and catch rate from the mid-nineteenth to the early twentieth centuries. We restricted our analyses to those communities where catches would have been primarily – in most cases exclusively – comprised of northern cod represented in the censuses: Ferryland, St. John's, and the communities of Conception, Trinity, Bonavista and Notre Dame Bays.

Despite a decline (6.5% to four percent) in the proportional representation of large boats (greater than thirty tons), the capacity to harvest northern cod doubled between 1845 and 1884 (harvest capacity is defined as the number of boats multiplied by the storage capacity of each boat) and remained constant thereafter (figure 4). Despite this doubling of harvest capacity, catches of northern cod between 1857 and 1891 decreased by more than thirty percent. (Note that a similar decline was not evident from an inspection of the total exports of cod from Newfoundland and Labrador, presumably because the latter included catches from the south and west coasts of Newfoundland.) Associated with the decline in northern cod catches was a greater than two-fold reduction in catch rate. It is

---

the "herring entirely abandoned the coast" (Gosling, *Labrador*, 425-426).

[96]*The Twillingate Sun*, 17 September 1887.

[97]*Ibid.*, 4, 18 and 25 August and 15 September 1888.

[98]See Hutchings and Myers, "Biological Collapse."

noteworthy that catch rates declined despite an overall increase in fishing effort: between 1845 and 1874, the number of nets and seines per unit of capacity increased ten-fold. The subsequent decline in the use of these gear types was concomitant with the introduction of the cod trap in the mid-1870s.

Thus, the rapid northward expansion of the Labrador fishery, which extended from the mid-1850s to the 1890s, was accompanied by declining catches, declining catch rates, increased fishing effort and increased harvesting capacity in the inshore northern cod fishery. These patterns are consistent with the hypothesis that expansion of the Labrador fishery was precipitated by declining inshore stocks of northern cod. If the size of inshore cod stocks had been stable during this period, the doubling in harvesting capacity and dramatic increases in fishing effort should have resulted in increased catches. Under these circumstances, catch rates might have been expected to increase initially (as fishermen began to harvest previously unexploited stocks) and then to decline to levels experienced previously.

It could be argued that factors unrelated to fishing (e.g., water temperature, salinity or changes in abundance of cod prey) were responsible for the apparent decline of inshore northern cod. The only long-term environmental data available for the nineteenth century are on ice-clearance recorded at Hopedale.[99] These data, which are significantly associated with water temperature, do not reveal any temporal trends through the 1800s.

---

[99]Hutchings and Myers, "What Can Be Learned," 2126-2146.

Figure 4
Temporal Changes in Various Metrics of Catch,
Fishing Capacity, Fishing Effort and Catch Rate, Ferryland,
St. John's, and the Outport Communities of Conception,
Trinity, Bonavista and Notre Dame Bays

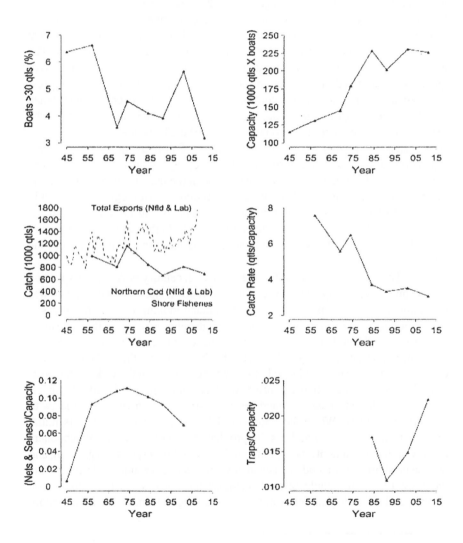

Source: Newfoundland, *Census of Newfoundland*, 1845-1911.

Conclusions

The deteriorating reputation of the quality of Labrador salt cod in European markets was a sign that expansion in the Labrador fishery was not without its problems. Going further afield meant longer handling of fish by floaters, who shipped it green in salt bulk back to the island for drying. Longer handling increased the possibility of damage, while those who tried to cure fish on the Labrador coast faced poorer conditions marked by damper weather. The floaters had, further, come to rely on cod traps, which yielded "miserable small fish" that cured poorly. Some advocated that the legislature should ban the traps. The merchants' practice of buying Labrador fish *tal qual* – for one price regardless of quality – exacerbated the problem of poor curing. In 1886, however, the press reported a better cure quality control system in the Labrador fishery when merchants paid according to grade, and fishers took greater care in handling the fish.[100] The problem of poor cure quality continued to plague the Labrador fishery. Aggravated by chaotic, cut-throat marketing by fish merchants, the declining reputation of Labrador cod ensured that the heyday of the Labrador industry would not be a long one. Credit over-extension in the industry contributed to the Bank Crash of 1894 that shocked the colonial economy. Although capital quickly reorganized the Labrador fishing industry, merchants and government did not address the problems of cure and competition. By 1920 the Labrador fishery was in marked decline, although a few floaters using long-liners hung on until the 1980s.[101]

Through the second half of the nineteenth century northward expansion had become a commonplace feature of the Labrador fishery. This expansion was not simply a matter of fishers from Newfoundland looking for new resources to fuel economic and social growth at home. The development of the floater fishery from the 1860s through the 1880s was instead the outcome of a growing ecological imbalance between outport people and the fishing resources they relied on almost completely. Intense interest in the Labrador fishery began in the wake of the post-1815 depression. The collapse in prices for Newfoundland-produced saltfish left a fairly new critical mass of island residents struggling to find new means to support themselves. While the Labrador fishery may have received its initial impetus from the need to address the problem of unused capacity of schooners employed in the seal hunt, its exploitation intensified in response to growing and persistent shortages of cod on island inshore fishing grounds relative to the increased effort by people through demographic factors, such as population growth, and technological innovation. Such shortages prompted more and more

---

[100]*The Twillingate Sun*, 20 February, 3 April and 2 October 1886.

[101]Cuff, "Labrador Fishery," 228-230.

people to try their luck in the less crowded waters of the Labrador coast. By the 1860s, however, similar conditions of scarcity relative to intensified effort began to plague the southern waters of the old Labrador industry. The herring fishery in those waters proved to be a fickle compensation. In consequence, some fishers adopted the floater strategy to draw into their range a wider field of resource exploitation. The northward expansion of the Labrador fishery, in combination with the introduction of more intensive harvesting gears, did allow periodic recovery and growth of salt-cod exports. Such cyclical patterns in exports hid an ecological problem: declining productivity in inshore waters.

Status and Potential of Historical and Ecological Studies on Russian Fisheries in the White and Barents Seas: The Case of the Atlantic Salmon (*Salmo Salar*)

Julia Lajus, Yaroslava Alekseeva,
Ruslan Davydov, Zoya Dmitrieva,
Alexei Kraikovski, Dmitry Lajus,
Vladimir Lapin, Vadim Mokievsky,
Alexei Yurchenko and Daniel Alexandrov

Abstract

In this paper we discuss the use of historical archival records, including direct statistics, tax records and cloistral accounts, to analyse the population dynamics of cod, herring and salmon in the Barents and White seas from the seventeenth to the twentieth century. The basic aim is to estimate the effect of social (fishing effort) and biological (climate changes and fluctuations in numbers) factors on marine fish populations. This is the first attempt to organise such historical research in Russia, despite the availability of large volumes of relevant and accessible primary source material. We include a case study on the salmon fisheries in the region. The first available data on salmon fisheries are from the beginning of the seventeenth century, although more complete statistics are not available until the 1870s. The data reveal that fewer salmon were caught at the beginning of the seventeenth century in the Onega River than in more modern times, given comparable effort. We also examine the relationship between catches, fishing effort and prices.

Introduction

In the framework of the HMAP programme our group is responsible for collecting and analysing statistical data relating to the most important commercial fish species: cod, herring and salmon. The data relates to one of Russia's oldest fisheries, which operated in the White and Barents Seas. Reasonably reliable data from the seventeenth century exist for these fisheries. For the historical and proto-statistical periods, data that indirectly reflect catches and fishing effort could be used, such as tax records, customs records, trade figures, and consumption of fish in the larger trading towns and monasteries. This is the first attempt to organise such historical research in Russia, despite the availability of a large volume of relevant and accessible primary source material sufficient to shed light on the question of ecological change.

Among the members of HMAP-Russia group are scholars working in the seventeenth-century economic history of the Russian North (Zoya Dmitrieva and Alexei Kraikovsky), experts in nineteenth-century Russian history (Vladimir Lapin and Ruslan Davydov), an ethnographer who specialises in northern culture (Alexei Yurchenko), three environmental historians (Daniel Alexandrov, Yaroslava Alekseeva and Julia Lajus) and two ecologists (Dmitry Lajus and Vadim Mokievsky). The collaboration of historians and biologists in the project is very important, not only for the research effort but also in terms of its management. It involves a dual system of responsibility in which historians are responsible for collecting and analysing data for each particular historical period, while biologists coordinate the collection of data on the main fish species and provide ecological analysis.

The general aim of our project is to describe the fluctuations in catches of the three main fish species and to analyse them against changing environmental and social variables. Through the incorporation of megafauna aspects into this analysis, we hope to reach an understanding of ecosystem changes in the White and Barents seas. During the first stage of the project we concentrated on compiling available historical data on catches and analysing the population dynamics of Atlantic salmon, a fish that was distributed widely in the basins of both seas and which formed the basis of the earliest and most valuable fishery in the region.

The other important focus of our study is the cod fishery that operated along the Murman coast of the Barents Sea from the sixteenth century through to the 1930s. The main goal of the historical study of these fisheries is to describe the fluctuations in migrations of the different species of fish (especially cod) for the period that preceded the era of over-fishing. We plan also to analyse the historical evidence for seasonal distribution of cod along the coast. Such research will allow us to connect modern fishery statistics with the data from the pre-statistical period and, by such means, to have a much better understanding of the migration patterns and population dynamics of cod and related species within a broader historical period.

Herring, a species characterised by a high tendency for fluctuations, could be a very sensitive indicator of environmental changes. All hypotheses explaining the decline of the White Sea herring population were based on, and verified by, recent fisheries statistics (drawn from the twentieth century in many cases; from the 1930s, modelling is based on data from the 1950s). A description of these fluctuations for the earlier period, when the fishing effort was probably more stable, may yield very valuable information for validating these hypotheses. We intend to consider the fluctuations of White Sea herring on a small geographical scale, analysing the data on different herring stocks separately. Although the lack of historical data for the Barents Sea herring fisheries restricted our research largely to White Sea herring, we hope to obtain some material on Barents Sea herring as well. Comparison of these data should allow us to define the periods of herring abundance in both areas, thus showing its possible correlation with general climatic fluctuations in the region.

By including megafauna in the picture we may ultimately gain a wider ecosystem perspective. In this respect we considered the Atlantic walrus as one of the most promising candidates for the possible reconstruction of population dynamics within a species. The number of walruses has fallen dramatically over the last 400 years, with a corresponding reduction in the area of its distribution. A study of the walrus, which has already been started in Russia, has the potential to organise and consolidate the work of the international group, together with researchers from other regions where the walrus is present. These combined efforts may provide a very good picture both historically and ecologically. As an additional indirect indicator of the size of fish populations and their migration patterns, statistics on the hunting of seals and white whales can be used. Seal hunting was always very important to the local economy, so the statistics on sealing are very rich.

In this paper we would like to give the historical and ecological background on which we based our research programme and to demonstrate some approaches and present some initial findings from the data on the Atlantic salmon fishery. All members of our group contributed to this publication and thus are listed as co-authors. The first-named author is the person who assumed the role of editor, and the last named is the project director. Other co-authors are listed in alphabetical order.

To facilitate collaboration between the different disciplinary and scholarly communities, it is crucial for the creation in Russia of a new research field: marine environmental history. It is hoped that one outcome of this research project will be to give a general impetus to such a development.

Russian Northern Sea Fisheries: General Background

The history of the fisheries in the Russian northern seas covers both the White Sea and the Barents Sea, and any discussion needs to consider both in tandem. This is due to the very close historical connections of these fisheries, and also because some species (Atlantic salmon, Atlantic herring and Greenlandic seal) are distributed in both these seas. Although some unique environmental and biological features characterise the White Sea as the inner sea, both are greatly influenced by the processes taking place in the northwest Atlantic.

When writing about the fisheries in the White and Barents seas we focus primarily on fishing by coastal dwellers, who are known in Russian as *Pomors*. This ethnic group formed over time and represented various nationalities, the major one being eastern Slavonic. The first records of these people reaching the White Sea coasts date back to the early eleventh century, and permanent Russian settlements were established by the close of the thirteenth.[1] The basis of the *Pomor* economy was fishing and marine hunting. The oldest fishery was for salmon in the rivers of the White Sea basin. As in the other Russian fishing regions, the main characteristic of its development was a gradual transition from freshwater fisheries to the fishing in the mouths of rivers and then in inshore waters.

The salmon has always played a special role in the Russian north. It was one of the main, if not the only, means of supporting the population. The colonisation of the area depended much on the abundance of salmon. From the very beginning salmon was one of the most important goods for trade and was the means of paying taxes and tithes. Through this trade, the new inhabitants of the White Sea coasts were able to maintain traditional Russian cultural practices and ways of living. The income generated from the salmon trade was used to buy wheat and ironware. On the whole, the salmon fisheries took priority for the *Pomors* over all other forms of earning a living.[2]

The White Sea salmon fishery was the main *Pomor* fishery up to the beginning of the sixteenth century, when the Barents Sea seasonal inshore long-line fishery began to develop. Cod (*Gadus morhua L.*) became an additional source of income, especially in export markets. The peculiarity of the Barents Sea fishery is its dependence on the migrations of fish. Cod, haddock and herring migrate to the Barents Sea from their spawning grounds near the Norwegian coasts. It is well-known that the intensive development of trawling from the beginning of the twentieth century, and especially in the 1950s, 1960s and 1970s, led to a significant decline of catches in cod and herring and to a decrease in the

---

[1]*Pomor* descends from the Russian word "more," which means "the sea." Thus, the *Pomors* are people who live near the sea (coastal dwellers). According to T.A. Bernshtam, *Pomory: formirovanie gruppy i sistema khoziaistva* (Leningrad, 1978), the name *Pomorie* in sixteenth-century documents was first used in relation to the uninhabited areas of the western part of the Barents Sea coast where Russian fishermen used to fish. Gradually, the name *Pomor* came to cover people who took part in the Barents Sea cod fishery. Later, all of the population of the White Sea coast took on the name.

[2]*Ibid.*; and S.F Ogorodnikov, *Ocherk istorii goroda Arkhangel'ska v torgovom i promyshlennom otnoshenii* (St. Petersburg, 1890).

capelin population.[3] Environmental, especially climatic, changes could also have been a factor in this decline. Russian data could provide information about changes in species abundance and composition due to climatic changes because Russian fisheries are situated in the eastern part of the Barents Sea and are based on migratory species that are very sensitive to climate.

The White Sea herring fisheries started to develop commercially later than the salmon and cod fisheries but played an important role in the local economy. It is based on two species: the Atlantic (*Clupea harengus L.*) and the Pacific herring (*C. pallasi Val.*). But while Atlantic herring comes to the White Sea only to feed and never spawns there, the Pacific herring, represented here by the sub-species *Clupea pallasi marisalbi*, lives there throughout its life cycle and is divided into many stocks.[4] The main trend in the White Sea herring fishery has been a long-term, if variable, decline in the catch over the last 200 years.[5] Almost nothing is known about this fishery before the middle of the eighteenth century, although some evidence indicates that it was developed on a commercial level at least one hundred years earlier.

Of all the marine mammals, it was the walrus that represented the primary commercial catch for the Russian population of *Pomorie*. There are numerous indications that, in the past, walrus distribution in the eastern Barents Sea was much wider than is the case today. Names of islands like Morzhovets and Morzhovy (*morzh* means walrus in Russian) in the southern part of the Barents Sea, and even islands in the White Sea, also point to possible distribution patterns in the past. The numbers of walrus tusks sold in the open market over the last centuries provide a more exact picture of the abundance of the species. Walrus tusks and tusk ivory, which were called "fish teeth" in the official papers of that period, were the most valuable product of the fisheries along the coasts of the White Sea. Every tenth walrus tusk was paid as a duty, and the collected tusks were sent directly to the Tsar's court. The significant changes in the ecosystems

---

[3]See, for example, V.O. Mokievsky "Biologicheskie resursy morei i presnykh vodoemov," in V. Yablokov (ed.), *Rossiskaia Arktika: na poroge katastrofy* (Moscow, 1996), 93-102; and V.O. Mokievsky and V.A. Spiridonov "Chto oznachaiut dlia Rossii ee morskie biologicheskie resursy," in N.N. Marfenin, N.N. Moiseev and S.A. Stepanov (eds.), *Rossia v okruzhaiuschem mire: 1999. (Analiticheskii ezhegodnik)* (Moscow, 1999), 119-134.

[4]D.L. Lajus, "What is the White Sea Herring *Clupea pallasi marisalbi* Berg, 1923?: A New Concept of the Population Structure," *Publicaciones especiales Instituto Espanol de Oceanografia*, XXI (1996), 221-230.

[5]The maximum known annual catch of the White Sea herring was 8260 metric tons in 1928 according to V.V. Kuznetsov, *Beloe more i biologicheskie osobennosti ego flory i fauny* (Moscow, 1960).

of the White and Barents seas can be explained by reasons of over-exploitation of populations and climatic changes, but only studies of historical data can give us a clear understanding of the processes taking place in the area.

Historical and Statistical Data on the Salmon Fisheries in the Russian North

The same ecological hypotheses of over-exploitation and climatic changes could be tested by looking at the dramatic decrease in the Atlantic salmon population in the Russian North. We have good reasons to believe that the Atlantic salmon is the most promising subject for an analysis of the population dynamics in this region over the long term for five reasons. First, there is the availability of sources over several centuries when salmon was abundant, and the salmon fishery was more advanced than other fisheries. Second, the wide distribution range of salmon in the Russian North permits us to make comparisons between different populations within the area. Third, the fisheries of the North are based mainly on spawning, when the fish enter a river mouth and move upstream. This migration is restricted to a limited period of time (summer and autumn), which makes river fisheries easier to study compared with those in the open sea. Fourth, we know that salmon catch sizes, even from river fisheries, are a good indicator of conditions in the North Atlantic ecosystem, especially in relation to climatic changes. Finally, technical improvements in fishing gear were not especially marked in salmon fishing. This makes comparisons of data sets from various periods feasible.

We currently have a data set on the salmon fisheries in the Onega, Severnaya Dvina, Varzuga, Mezen' and Ponoy rivers for a number of years during the seventeenth century. The salmon fisheries in the eighteenth century need to be studied carefully on the basis of archival records of private monopolies, which controlled them for most part of the century. This form of ownership had a significant influence on the development of the fisheries in general and the catch sizes of salmon in particular. The most complete statistical data on the salmon fisheries exist for the period from 1875 to 1915 in the records of the Statistical Committee of Arkhangelsk Province. Thus, the significant part of our analysis is based on these data, with the addition of more recent evidence taken from statistical and scientific sources.

Biological Features of the Salmon as a Component of Marine Ecosystems

The Atlantic salmon (*Salmo salar L.*) is a migratory species which feeds in the open North Atlantic and comes back to its home rivers to spawn. The salmon is distributed in the western Atlantic from northern Québec in Canada to Connecticut in the USA, and in the eastern Atlantic, from the Arctic Circle to Portugal. The Barents and White seas comprise the largest area of distribution of the Atlantic salmon in Russia. Indeed, after the depletion of the Baltic Sea population, it is the only area inhabited by wild salmon. The coastlines of these two seas are heavily indented and are fed by numerous small and large rivers which are very suitable for salmon spawning.

The maximum size of the male Atlantic salmon is 150 cm.; females may reach 120 cm.; and maximum weight is up to 39 kg. Their life span is as high as thirteen years. The smolts live in the river for from one to six years. They then go to sea, taking long migratory routes far from their home rivers. Several years later (usually one to four), the adults return and may spawn up to five times. Adults in fresh water which are approaching the reproductive stage do not feed. Juveniles (smolt) feed on molluscs, crustaceans, insects and fish; adults at sea feed on squids, shrimps and fish.

It is well known that the size of the salmon population depends to a large extent on the conditions during the marine period of its life. Moreover, the current degradation of many populations, which started in 1986 and has affected all countries, has not been abated by a reduction in the number of fisheries and other measures aimed at restoring salmon numbers. This has led to the conclusion that the main cause of this decline has been due to high mortality during the marine stage of its life.[6] The mortality rate at sea can be very high and variable; according to some data, it can exceed the mortality rate during the freshwater stage by as much as three times.[7]

It has been established that the salmon is very susceptible to changes in sea temperatures and can modify its migratory routes depending upon the temperature of the upper water layer.[8] During the marine stage of its life history, the optimum temperature ranges from four to eight degrees Centigrade. The start of migrations, and the time of coming back to fresh water, also depend on water temperatures.[9] The growth of the Atlantic salmon takes place mainly at sea, and consequently its population is dependent on the status of the marine ecosystem and reflects its condition. Therefore, there are grounds to conclude that fluctuations

---

[6]D.G. Reddin, "Osobennosti morskogo perioda zhizni atlanticheskogo lososia," in R.V. Kazakov (ed.), *Atlanticheskii losos'* (St. Petersburg, 1998), 266-286.

[7]E.M.P. Chadwick and R.G. Randall, "Production of Atlantic Salmon (*Salmo salar*) in Fresh Water and at Sea at High and Low Densities," *Ecology of Freshwater Fish*, II (1993), 67-72.

[8]D.G. Reddin and W.M. Shearer, "Sea-surface Temperature and Distribution of Atlantic Salmon in the Northwest Atlantic Ocean," *in American Fisheries Society Symposium on Common Srategies in Anadromous/Catadromous Fishes, Part 1* (Bethesda, MD, 1987), 262-275.

[9]D.G. Reddin and K.D. Friedland, "Marine Environmental Factors Influencing the Movement and Survival of Atlantic Salmon," in D. Mills (ed.), *Salmon in the Sea and New Enhancement Strategies* (London, 1993), 79-103.

in salmon numbers, estimated by catches in fresh water, reflect the population dynamics during its marine stage and can provide important information on the status of the marine ecosystem.

It has been noted many times in the scientific literature that there is a connection between catches and population dynamics of salmon throughout all areas of its distribution. In 1935, for example, a famous Russian ichthyologist, Lev Berg, noted that fluctuations in salmon catches were taking place simulta- neously in various parts of western Europe, Russia and Northern America.[10] He noted, however, that this phenomenon required further investigation. According to L.B. Kliashtorin, an analysis of the long-term dynamics of Atlantic salmon catches during the last seventy years reveals commonalities with long-term herring and cod catches.[11] This apparently is evidence for the existence of common long-term patterns in general fish productivity in the North Atlantic. The common patterns in the population dynamics of the salmon from both America and Europe also point to the dominant role of the marine stage in determining population size. Global temperature seems to be an essential factor in this process.

For the long-term success of further studies in the history of population dynamics, it is important that analyses be extended to include the entire area of salmon distribution. This is possible only with the formation of an international group.

General Description of Salmon Fishery and Fishing Gear

Salmon fisheries in the Russian North were seasonal, lasting from June to November. As a rule, the main interest was pre-spawning fish. The most valuable catch for fishermen was the large pre-spawning autumn salmon, which go upstream from August until the rivers freeze; spend the winter in the river; and spawn the following autumn.

Various kinds of fishing gear were used to catch salmon. It appears that the earliest was the weir (*zabor* in Russian), a dam across the river with traps. Until the seventeenth century, river fisheries predominated but later, when the gear became more sophisticated, the emphasis started to shift towards the marine coastal fisheries. By the twentieth century, more than twenty different kinds of fishing gear were used in the river and marine fisheries. The development of

---

[10]L.S. Berg, "Materialy po biologii siomgi. Obzor rabot po issledovaniyu siomgi, provedionnykh v 1930-34 g. Vsesoyuznym institutom osiornogo y rechnogo rybnogo khoziaystva," in *Izvestiia VNIORKh, 20-Semga, eio biologiia i promysel* (Leningrad, 1935), 3-113.

[11]L.B. Kliashtorin, "Klimat i dolgoperiodnye fluktuatsii zapasov atlanticheskogo lososia," in *Atlanticheskii losos' (biologiia, okhrana, vosproizvodstvo)* (Petrozavodsk, 2000), 27.

marine fisheries started later in Russia than elsewhere due to the abundance of river salmon; marine fisheries were less profitable and more labour intensive.

Figure 1
Construction of a Weir on the River Ponoy

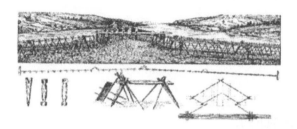

Source: *Risunki k issledovaniiu rybnykh i zverinykh promyslov na Belom i Ledovitom moriakh* (St. Petersburg, 1863).

The *zabor* retained its most important design features for centuries (see figure 1). It was comprised of branches and twigs, and the rivers were dammed at the places that formed part of the salmon's migratory spawning routes. Several fish traps were built into the body of the weir. Some of the approaching fish could escape upstream through gaps in the weir or swim around it if the weir had not closed off the river completely. After heavy rains, the subsequent floods could also facilitate escape. Drifting ice in the autumn could completely destroy a weir, and ice would often destroy the gear even before the salmon's migratory passage had been completed. At times, a weir could last a winter but be destroyed by the spring floods or demolished by the peasants themselves.

A drift net called the *poezd* was another popular piece of fishing gear (see figure 2). It mainly was used on the large rivers. The *poezd* was actually a sack with a very broad mouth which fishermen stretched between two boats and carried just above the river bottom. Its length was four to six metres, with widths of between six and 8.8 metres. Each boat usually carried two fishermen, one of whom rowed while the other held the gear. When the fishermen felt the push of the trapped fish, they moved the boats together, hauled the *poezd* out of the water, took out the fish, and then continued fishing.

Figure 2
Fishing of Salmon with a *Poezd* and the Construction of this Gear

Source: See figure 1.

Figure 3
Salmon Fishing with a *Garva* and the Construction of this Gear

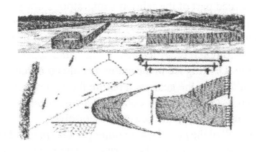

Source: See figure 1.

Table 1

Collections of the "Tenth Fish" According to the Account Books of Novgorodskaja Chetvert

| # | Year of Record | Year Caught | Tenth Fish | | | | | | Source |
| | | | Dvina | | | Onega | | | |
			Fish	Poods (16.38 kg)	Average Weight (poods)	Fish	Poods (16.38 kg)	Average Weight (poods)	
1	1614/15	1615	898	149	0.165	176	70.25	0.4	RIB, vol. 28 Prihodo-rashodnyje knigi Moskovskih prikazov. Vol. 1. M., 1912.
2	1619/20	1619	1335	260	0.19	100	42	0.42	Prihodo-rashodnyje knigi Moskovskih prikazov 1619-1621 rr. M. 1987
3	1620/21	1619	ND*	ND	ND	96	47	0.49	Ibid.
4	1625/26	1624	853**	168	0.2	ND	ND	ND	RGADA, f. 137, 90. 1, Novgorod, d. 15
5	1625/26	1625	742	123	0.165	129	62	0.48	Ibid.
6	1642/43	1642	577	187	0.32	61	41.5	0.68	Ibid., d. 28
7	1686/87	1686	390	77	0.2	ND	ND	ND	Ibid., d. 117
8	1688/89	1688	390	106.5	0.27	71	ND	ND	Ibid., d. 123

Note: *ND - no data. In the credit book of the Novgorodskaja chetvert' for 1620/1621, it is recorded that the tenth salmon from Dvina was not delivered before the compilation of the book and "is to be declared and delivered in the year 130 [1621/1622]."

**In the credit book for the year 1625/1626 the tenth fish from Dvina for the previous year is recorded.

Sources: See table.

Fishing with the so-called *garva* has a long history.[12] The *garva* was a kind of semi-permanent net (see figure 3). Its principal design remained the same over the years except for improvements in the quality of the mesh. The net was set from the shore towards the open sea or the middle of the river. Its outward-facing side ended with a trap of a somewhat complicated design. Fish entered at the shoreline or riverbank opening and continued to swim along the net until it finally reached the trap at the end. Quite often fishermen used harpoons, especially to catch spawning or newly-spawned fish. For example, on the Niva River harpooning was practiced until 1925. Annual catches with this method reached five tons.[13]

Salmon Catches According to the Account Books of *Novgorodskaia chetvert*

Thus far no quantitative data on salmon fisheries prior to the seventeenth century have been discovered. The evidence concerning the seventeenth century can be found in the account books of various government departments (*prikazes*) and monasteries.

The *prikazes* were central government bodies of the Moscow State, and their remit included collecting direct taxes and quitrents. The *prikaz* (department) of the *Novgorodskaya Chetvert'* (Novgorod area) was in charge of collecting payments from the Russian North. Data on taxes and other duties were registered in the credit books. Among numerous items of income in the books of this *prikaz* are data on the collection of the "tenth fish," i.e., the fisheries duty on salmon catches. It is important to note that the "tenth fish" mentioned here indeed meant one-tenth of actual catches. The historical source notes that "they collect for the Tzar...one tenth of the red fish called salmon, depending on how much there has been caught every year."[14]

At times the "tenth fish" was collected in the monetary equivalent. For example, in the sixteenth century in the rural district (*volost*) of Varzuga, peasant

---

[12]A. Yurchenko, "Morskie orudiia lova semgi pomorov v kontse XIX – nachale XX veka," *Nauka i biznes na Murmane*, II (2000), 31-36.

[13]V.L. Issatchenko, "Issledovaniia semgi, eio promysla i vyiasnenie v rekakh severa mest, prigodnykh dlia provedeniia meropriiatii po iskusstvennomu eio razvedeniiu," *Izvestiia of Leningrad Research Ichthyological Institute*, XIII, No. 2 (1931), 31-59.

[14]Russian State Archives of the Ancient Statements (RGADA), f. 137, Books of *Boyars* and Towns, op. 1, Novgorod, d. 15.1, 310.

fishermen were paying 140 *roubles* per year instead of the "tenth fish."[15] Nevertheless, it was evident that the government could, at any moment, demand to receive the "tenth fish" in kind. In 1615 the same *volost* delivered 3282 salmon to a total weight of 360 *poods* or about 5.9 metric tons (1 *pood*=16.38 kgs.).[16]

Records concerning the "tenth fish" in the credit books of the *Novgorodskaya Chetvert'* exist until 1690. After this year there are no longer data on fish in these books. Perhaps this was due to changes in the place of fish delivery. It is possible that from 1690 this task was made the responsibility of the Food Office (*Kormovoi Dvorets*) but since its archives were lost, this cannot be known with any certainty.

The existing sources enable us to reconstruct the process of collecting the "tenth fish." As a rule, this was collected by tax collectors (*tseloval'nik*) elected by the local peasants. Every fisherman had to show his catch to the *tseloval'nik*, who recorded the information in a special book on "tenth fish collecting."[17] One-tenth of the catch was taken out. At times, these books also provided information on the totals. For example, in 1682 in the ancestral domains (*votchina*) of the Krestny Onezhsky monastery were caught "in the *volost* of Podporozhie sixty-eight fish, in the *volost* of Porog fifty-six fish, [and] in the *volost* of Ustie forty-six fish. Total, 170 fish. Taken out, seventeen tenth fish, and four newly caught fish."[18] As a rule, the tax collector verified the records with his signature. If he was illiterate, a local Orthodox priest signed in his place.[19] If the tax collector took every tenth fish from the fisheries of the monastery, he gave a special receipt. For example, in 1637 the Ponoy customs tax collector, Piatoi Alexeev, received from the Ekonga fisheries of Antonievo-Siiski monastery thirty-three *roubles* and eleven *altyns* in cash instead of the tenth fish (that would have amounted to 3030 salmon). The person who made the payment was given a receipt.[20]

---

[15]A.I. Kopanev, "Nezemledelcheskaia volost v 16–17," in *Krestianstvo i klassovaia bor'ba v Rossii. Sbornik statei pamiati I.I. Smirnova* (Leningrad, 1967), 180-193.

[16]*Prikhodo-raskhodnye knigi moskovskikh prikazov 1619-1621 gg.* (Moscow, 1983), 71.

[17]These sources were found in the archive of the Krestny Onezhsky monastery. See, for example, RGADA, f. 1195, op. 1, d. 190, 204 and others.

[18]*Ibid.*, d. 334. 1, 140.

[19]See, for example, *ibid.*, d. 323. 1, one back, two back, three back, four back and five back.

[20]*Sbornik materialov po istorii Kolskogo poluostrova* (Leningrad, 1930).

Next, the collected fish were sent to Moscow, where the officials of the *prikaz* of *Novgorodskaya Chetvert'* kept credit books according to the districts (*uezd*); the amount of the "tenth fish" tax was also recorded in this way. It is important to note that in the Russian North, districts as a rule were formed around separate rivers. Therefore, the aggregate of the "tenth fish" tax collected in a particular year from a particular district indicated the level of annual salmon catches from a particular river (including tributaries) within well-defined boundaries. The credit books of the *prikaz* of *Novgorodskaya Chetvert'* show clearly data from the Dvina River (Dvinskoy *uezd*) and the Onega River (Turchasovo and Kargopol'). For example, there is this record from 1620:

> Year 128 [1620], January 3. According to the message from Dvina from governor (*voevoda*) prince Andrei Hilkov and clerk (*diak*) Semeika Zelenyi, the company commander of the riflemen (sotnik streletskii) Andreika Zahariev and Dvina fish tax-collector Osipko Stepanov delivered the Dvina tenth fish – salmon, collected in the year 127, 1335 fish, weighing 260 *poods* [or about 4.25 metric tons].[21]

After being recorded in the *prikaz* of *Novgorodskaya Chetvert'* the fish were sent to the office of the Great Palace (*prikaz Bol'shogo Dvortsa*), and later to the Food Office.[22] It is important for our analysis that the tax collector, who was responsible for the delivery of the tenth salmon to Moscow, ensured that the actual weight of the salmon delivered matched the weight stated in the message from the governor and the clerk, since he had to cover any shortfall. For example, Dvina fish tax collector Maksim Ogryzkin in 1681 was forced to pay for some missing salmon.[23]

Table 1 shows the results of collecting the "tenth fish" at the Dvina and Onega rivers and is based on data from the credit books of the *prikaz* of *Novgorodskaya Chetvert'* for the seventeenth century. As the table shows, the catches on both rivers fluctuated throughout the century. Moreover, catches from the Onega River became progressively poorer. Analysing this tendency, we have to take into consideration at least three factors: the possible shortcomings of the accounting system; an increase in the number of fishermen paying their "tenth fish" in money instead of kind; and the changes in fishing capacity of the gear being used.

---

[21]*Prikhodo-raskhodnye*, 66.

[22]*Ibid.*

[23]RGADA, f. 159, op. 3, d. 1318, l, 1-10.

Apart from these hypothetical reasons, the decrease in catches could be caused by several other variables. First, it is possible that human exploitation caused a decrease in the salmon population. This assumption almost certainly is true. In the seventeenth century, the methods of resource exploitation were predatory. In 1676, for example, the metropolitan of Novgorod sent a charter to the Krestny Onezhsky monastery in which he explicitly forbade it to dam the Onega River and pointed out that a passage should be left for the salmon.[24] Nevertheless, it is obvious that exploitation alone could not cause a significant decrease in the catch. As will be shown below, both the Onega and Dvina rivers remained the most important salmon fisheries until the middle of the twentieth century, demonstrating that their salmon populations were able to withstand even more significant pressures.

Second, it is possible that the numbers of fishermen were also decreasing. In order to check this hypothesis we need to study thoroughly data on the human population in the Russian North. A general assessment was made by A.I. Kopanev. According to the cadasters of the seventeenth century, the *Pomor* population declined significantly between 1620 and 1640 and was not restored until 1670. At the Dvina *uezd*, the population decreased by 36.8% between the 1620s and 1640s. While by the 1670s population grew by twenty percent, the absolute numbers were still below the corresponding levels of the 1620s. At the Mezen' River, between the 1620s and 1670s the decline in the population was seventy-five percent.[25]

Third, a significant fall in the average annual temperature, which took place in the second half of the seventeenth century, could also have played a role in the observed decline in salmon catches by directly affecting the salmon population and/or reducing significantly the fishing effort due to the shortened fishing season. It is well known that in northern Europe during this period the average annual temperature was lower compared with previous years. Temperature profiles of the Svalbard ice cores tells us that much of seventeenth century was affected by cold weather: the mean summer temperature was approximately 1.5° C. lower than the mean temperature of the first half of the twentieth century. There was a short spell of warmer weather that lasted from 1625 to 1645. After this, the coldest period in the seventeenth century began, lasting until 1680.[26] For

---

[24]*Ibid.*, f. 1195, op. 2, d. 198.

[25]A.L. Shapiro, *Agrarnaia istoria Severo-Zapada Rossii XVII veka (naselenie, zemlevladenie, zemlepol'zovanie)* (Leningrad, 1989).

[26]Louwrens Hacquebord, "The Hunting of the Greenland Right Whale in Svalbard, Its Interaction with Climate and Its Impact on the Marine Ecosystem," *Polar Research*, XVIII, No. 2 (1999), 375-382.

the European part of Russia A.V. Dulov has produced the following evidence from historical sources: for the period between 1551 and 1600, weather conditions were mentioned in seventy-one cases, out of which cold winters were mentioned seven times and mild ones five times. During the following fifty years, cold winters were mentioned in twelve cases and mild ones only twice. Between 1651 and 1700, on nineteen occasions winters were characterised as unusually cold, whereas mild winters were mentioned three times.[27] An analysis of data in the old Russian chronicles suggests that the cold period affected the Russian North as well. For example, unusually bitter frosts, which were extraordinary even for this region, were noted in 1656-1663, 1671, 1679, 1680, 1686, 1689, 1690, 1692, 1695, 1696 and 1697.[28]

It is significant that there is archaeological evidence to show that in the seventeenth century there was also a major change in the distribution range of the Atlantic salmon. There was, for instance, a large increase of the salmon in New England (the southwestern part of its range). This change could be connected with cooling which made conditions there more optimal for the salmon.[29] It would be worthwhile to find any further evidence of the depletion of the salmon population during the same period in the northeastern part of its range, which is actually the area of our research. This evidence could be used to illustrate the general picture of salmon distribution over the cold period.

Another characteristic feature of the data in table 1 is the high variability in average fish weight. As the table shows, Dvina salmon was on average smaller than salmon from Onega. It is notable that average fish weight could fluctuate annually. For example, in 1642 salmon was larger overall than in other years in both rivers. This can hardly have been a mistake in the accounts, as the increase in fish size was noted in the Dvina River as well as in the Onega. Additionally, the collection of the "tenth fish" at both rivers was conducted by different people, thus lessening the probability of error. For a better analysis of this issue it would be necessary to study thoroughly the time periods when the fisheries took place (some credit books also contain records showing periods of activity). It is well

---

[27]See A.V. Dulov, *Geograficheskaia sreda i istoriia Rossii k. XV – ser. XIX v.* (Moscow, 1983), 14-30. Special attention should be paid to table 2, which gives a summary and preliminary assessment of climatic conditions for a period of about 500 years.

[28]E.P. Borisenkov and V.M. Pasetsky, *Tysiacheletniaia letopis' neobychainykh iavlenii prirody* (Moscow, 1988), esp. 335-345.

[29]Catherine C. Carlson, "The (In)Significance of Atlantic Salmon," *Federal Archaeology*, VIII, Nos. 3/4 (1996), 22-30; and Carlson "'Where's the Salmon?' A Reevaluation of the Role of Anadromous Fisheries to Aboriginal New England," in George Nicholas (ed.), *Holocene Human Ecology in Northeastern North America* (New York, 1988), 47-80.

known that average salmon weight can be very different depending on the time of entry into the river. At the beginning of the spawning season mainly smaller fish (*tinda*) entered the rivers, whereas at the beginning of August the size of fish was much larger.

Catches at the Onega River According to the Cloistral Records

Apart from data from the credit books of the *prikazes*, additional information can be obtained from the credit and debit books of the monasteries which owned the fisheries at the Onega and Dvina rivers (Krestny Onezhsky, Nikolo-Karlsky, Antonievo-Siisky, Kirillo-Belozersky and others). Unfortunately, the possibility to quantify catches during the sixteenth and seventeenth centuries, let alone for an earlier period, is rather poor due to the state of the historical sources. The following example is illustrative. In 1704 the government of Peter the Great decided to reform the tax system for all cloistral fisheries and asked for the necessary information, including records of catches. The majority of the monasteries of Vologda eparchy were unable to provide these types of data. A typical response was submitted by the elders of the Spaso-Prilutskii monastery: "And how much fish is caught per year, has never been recorded in the monastery, and no sources to learn from."[30] For that reason, we regard it as a great success that we have already found a number of cloistral documents that registered catches in the archives.

The materials from the Krestny Onezhsky monastery stand out for their information content. This monastery, besides having its own fisheries, also had the right to control collecting the "tenth fish" from all fisheries in the vicinity. This makes it possible to consider in detail the data recorded at the *prikazes* and to compare them with the credit books of the elected tax collectors, copies of which were kept in the monastery's archives. For 1688, we have comparable data from these two sources. According to the *prikaz* data, in 1688 there were seventy-one "tenth fish" from the Onega River; the total catch therefore was 710. In the books of the tax collectors we find records for about 622 fish, including catches from fishing grounds and weirs which belonged to the monastery. The discrepancy is less than thirteen percent. This evidence is important because it shows clearly that catches from places under the authority of the Krestny Onezhsky monastery constituted the major part of all catches from the Onega River. Therefore, data for other years can be used to estimate total catches in this area.

The archive of the Krestny Onezhsky monastery also contains a number of books of tax collectors from 1669 to 1704. The catches were recorded separately for each fishery: Podporozhskaya *volost*, Porozhskaya *volost* and "fishing grounds." Records in these books were kept in a specific manner: they

---

[30]State Archives of Vologda Region (GAVO), f. 496, op. 1, d. 6,1, 36 ob.

listed by name only those fishermen who were "showing the salmon in the catches." Fishermen who did not have salmon catches were not registered. This means that the number of fishermen mentioned in the credit books does not reflect the total fishing effort, as it can be assumed that the entire male population of the area was engaged in the salmon fisheries. In order to estimate total fishing effort we need to turn to other sources, in particular the cadasters. It should be pointed out that the records on catches by peasants were more detailed and thorough than those by monks. These latter records can be found in the books of the tax collectors of the Podporozhskaya *volost* for only six years (1685-1689 and 1691).

This material forms the basis of table 2, which shows for those six years the maximum amount of fish caught in the Podporozhskaya *volost* (1474 by peasants, and 1303 by the monastery). The catches from the Porozhskaya *volost* were about one-third this number (842). Even fewer fish were caught at the fishing grounds (556, or 13.3% of the total). Unfortunately, it is impossible to discern any direct correlation between the amount of fish caught and the type of gear used. For example, in 1699 at the Ustianskaya *volost* the owners of two neighbouring fishing grounds (Tikhmanskaya and Okoemovskaya) were using the same type of fishing gear, but the catch sizes were markedly different (twenty-two and six, respectively).[31]

In various years the catch sizes were different as well. Thus, in 1686 the number of the salmon caught was 182, while in 1689 it was 1380. Apart from a general tendency for a decrease in the catch, there was also a trend for a decline in catch sizes in every area. There were only two exceptions to this tendency. In 1687 the total catch was 246; out of this number, seven (2.8%) were caught at the "fishing grounds" and 113 (45.9%) in the cloistral fisheries. In 1689 in Podporozhskaya *volost* peasants caught almost a half the total number of salmon for the year (46.2%), while peasants from Porozhskaya *volost* caught only 12.9%. Yet in other years peasants from the latter *volost* usually caught about one-quarter of the total. On average, one fisherman from Porozhskaya *volost* caught six salmon, and a fisherman from Podporozhskaya *volost* landed 8.4 salmon. From one fishing ground an average catch was 11.8, or about 1.5 times greater than a catch made by *poezd*. Table 2 also shows that the number of fish caught corresponds to the number of fishermen registered in the books of the tax collectors as well as to the fishing area. On the whole, the fisheries in Podporozhskaya *volost* were more productive.

Analysis of the Salmon Fisheries, 1875-1915

The main source for the quantitative data on catches, number of fishermen and value of catches was the Statistical Committee of the Arkhangelsk Province, which

---

[31]RGADA, f. 1195, op. 1, d. 158, l, 3 back.

collected the most complete and consistent data on the salmon fisheries from 1875 to 1915.[32] Between 1903 and 1913 general statistical data that separated the White and Barents basins were also published in the Bulletin of the International Council for the Exploration of the Sea (ICES).[33] These data were based on Statistical Committee as well. In addition, we used the Russian State Archives of Economy (RGAE) and data from scientific publications which, however, are not numerous for this period.

Until the 1870s the sources provide relatively little hard statistical data on the fisheries. Even so, from these data it can be assumed that throughout the whole of the nineteenth century catches and the amount of fishing gear were both increasing. Yet despite an increase in gear used, the intensity of fishing was relatively low until the beginning of the twentieth century. Indeed, until the 1920s fishing techniques were rather primitive, and the type of gear used required much time and effort. The population was low compared to the fishing area, so it is unlikely that it would have had a negative impact on most of the salmon populations. For instance, at the mouth of the Kem' River, where the number of weirs was greater than elsewhere, and at the large Onezhsky *zabor*, catches were quite stable and even showed a tendency to increase.

Contemporaries often argued that the number of salmon was decreasing and considered over-fishing of spawning and post-spawning fish to be a contributory factor; in particular, they singled out the use of weirs.[34] From as early as the end of the seventeenth century, attempts were made to prohibit them, but these were rarely, if ever, successful.[35] Over time, the number of weirs

---

[32]*Otchety Arkhangel'skogo Statisticheskogo Komiteta*, 1876: (Arkhangelsk, 1876); 1878: (Arkhangelsk, 1880); 1882: (Arkhangelsk, 1884); 1883: (Arkhangelsk, 1885); 1884: (Arkhangelsk, 1887); 1885: (Arkhangelsk, 1887); 1886: (Arkhangelsk, 1888); 1887: (Arkhangelsk, 1889); 1890: (Arkhangelsk, 1892); 1891: (Arkhangelsk, 1893); 1892: (Arkhangelsk, 1893); 1894: (Arkhangelsk, 1895); 1895: (Arkhangelsk, 1896); 1896: (Arkhangelsk, 1897); 1897: (Arkhangelsk, 1898); 1899: (Arkhangelsk, 1901); 1902: (Arkhangelsk, 1904); 1903: (Arkhangelsk, 1904); 1904: (Arkhangelsk, 1906); 1905: (Arkhangelsk, 1908); and *Obzory Arkhangel'skoi gubernii* (Archangelsk, 1905-1915).

[33]*Bulletin Statistique des Peches Maritimes des pays du Nord de l'Europe*, I-VII (Copenhagen, 1906-1914).

[34]"Vtoroi (II) s'ezd rybopromyshlennikov Pomorsko – Murmanskogo paiona v Sumposade," *Vestnik rybopromyshlennosti* (1915), 7-8 and 467; and R.P. Yakobson, "Otchet po obsledovaniiu rybolovnykh tonei na rekakh Kemi i Vyge osen'iu 1911 goda," *Materialy k poznaniiu russkogo rybolovstva*, II, No. 9 (1914), 39-40.

[35]Yakobson, "Otchet po obsledovaniiu," 10.

decreased, but this was due not to prohibitions by the administration but rather to an expanding timber industry: weirs impeded timber-rafting, which was vital to the transport of logs. Conversely, from the 1890s to the 1930s timber-rafting was considered to be the main cause of the decrease – or sometimes the complete disappearance – of salmon stocks in many rivers.[36]

As a rule, statistical data are provided by administrative region, the largest of which was Arkhangelsk province (*guberniia*). Each province was divided into districts (*uezd*) and rural districts (*volost*), which were concentrated around one or two large settlements. But the boundaries and the lesser regions of Arkhangelsk province changed over time. For example, in 1891 one of the main areas of the salmon fisheries, Pechorsky *uezd*, was classed as an administrative unit distinct from Mezensky *uezd*. Thereafter, statistical data on the Pechorsky *uezd* began to be recorded separately as well.[37] As a result, in studying the history of the salmon fisheries it is crucial to identify the specific fishing areas within the different administrative regions using contemporary maps and reference books.

Figure 4
Catch Sizes of Atlantic Salmon (Metric Tonnes) in the Russian North,
1875-1915

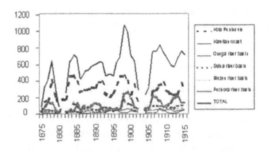

Source: See text.

---

[36]For example, the Semzha, Von'ga, Keret', Kem', Luven'ga, Kolvitsa and Kovda rivers according to V.V. Nikolsky, *Byt i promysly naseleniia zapadnogo poberezh'ia Belogo moriia (Soroki-Kandalaksha). Po materialam issledovaniia letom 1921 goda* (Moscow, 1927), 169. See also Kuznetsov, *Beloe more i biologicheskie osobennosti*, 27.

[37]*Administrativno-territorial'noe delenie Arkhangel'skoi gubernii i oblasti v 18-20th vekakh. Spravochnik* (Arkhangelsk, 1997).

Table 2

Catches of Salmon in the Patrimony of Krestnyi Onezhskii Monastery in the Seventeenth, Century, According to the Books of the Fish Tax Collectors

#	Year	Porozhskaja volost (1)		Podporozhskaja volost (2)		Fishing Grounds (3)		For the Monastery	Total		Fish per Fisherman		
		Fisher-men	Catch (fish)	Fisher-men	Catch (fish)	Fisher-men (fishing grounds)	Catch (fish)		Fisher-men	Catch (fish)	1	2	3
1	1685	31	199	35	226	6	90	210	72	725	6.4	6.5	15
2	1686	18	64	10	36	7	32	50	35	182	3.6	3.6	4.6
3	1687	20	85	12	41	3	7	113	35	246	4.3	3.4	2.3
4	1688	30	139	24	143	11	115	225	65	622	4.6	6	11
5	1689	24	178	55	638	11	169	395	90	1380	7.4	11.6	15
6	1691	18	177	39	390	9	143	310	66	1020	9.8	10	16
7	Total	141	842	175	1474	47	556	1303	363	4175	6	8.4	12

Source: RGADA, f. 1195, op. 1, d. 364, 383, 399, 414-415, 434 and 458-459.

We divided the distribution of Atlantic salmon in the Russian North into the following regions: rivers and marine fishing grounds of the Kola Peninsula; rivers and marine fishing grounds of the Karelia coast of the White Sea; rivers and marine fishing grounds of the eastern coast of Onega Bay of the White Sea, including the Onega River and its tributaries; rivers and marine fishing grounds of the coasts of the Dvina Bay, up to the river Zimnyiaa Zolotitsa, including the River Severnaya Dvina and its tributaries; rivers and marine fishing grounds of the Zimnyi coast, starting from the Megra River, including Mezen' Bay and River, up to the Cheshskaya Bight; and the Pechora River and its tributaries. There are two reasons for dividing the area in this way. First, this division corresponds roughly to the administrative regions and accordingly facilitates the use of the statistical data. The administrative regions were more or less built around large rivers (Onega, Severnaya Dvina, Mezen', Pechora) and therefore also reflect the distribution of the main salmon populations.

The dynamics of different catches in each of these regions display their own patterns, but at the same time they show a close resemblance (see figure 4). A statistical analysis has yielded correlation coefficients between catches in various *uezd* during the period 1875-1915 of between 0.18 to 0.80; in nearly all cases they were statistically significant.

Feasibility of Using Indirect Data to Analyse Salmon Population Dynamics

The possibility to interpret indirect data (prices, number of fisherman, etc.) is of crucial importance for historical studies because direct data on catches are not always available. Indirect data were recorded much more frequently in the past than were direct data. For example, there is an interesting set of data on the salmon fisheries at the Pechora River for the middle of the nineteenth century.

> The summer catches of salmon…in poor years reach up to 4800 *poods*…in good years on the average 23200 *poods*…In 1852 the catches were especially good, and the price was 1 *rouble* for 1 *pood*; in 1853 the price was 1.50 *roubles*; in 1854 from 1.70 to 2.60 *roubles*; in 1855 (poor catches) – 3 and 3.50 *roubles*; in 1865 – 4 *roubles*; in 1857 (very poor catches) – 5.15 *roubles*; in 1858 – 4 *roubles*; in 1859 – 5.43 *roubles* and 6 *roubles* in silver for 1 *pood*."[38]

These data show that the price of salmon varied depending on the catches and that the relationships between these two variables was close to inverse.

---

[38]G. Terentiev, "Sel'sky byt i promyshlennost' pustozertsev," *Arkhangelskie Gubernskie Vedomosti*, 26 November 1860.

For the entire data set we tried to establish possible correlations between different evidence from the Statistical Committee. This allowed us to reveal some significant patterns. Between 1875 and 1899 the number of fishermen almost doubled, from 4300 to 9000 men (only years with separate data on salmon fisheries were considered). The size of catches also doubled, as did salmon prices. There is a definite correlation between the number of fishermen and fish prices (see figure 5). Most probably, it was an increase in price that brought about the consequent increase in the number of fishermen. At the same time, no substantial increase in the average income of fishermen was noted.

Figure 5
Catch Sizes, Prices and Number of Fishermen in the Salmon Fisheries,
1875-1915

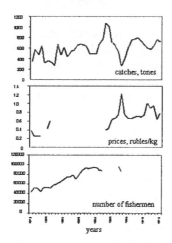

Source: See text.

The overall correlation between size of catches and number of fisherman is quite weak ($r = 0.36$). This shows that the size of the catch is influenced by a number of factors not related directly to the number of fishermen. One is obviously the issue of salmon population dynamics. The price of salmon and the size of the catch showed a negative correlation, although the absolute value was low ($r = -0.17$). Evidently, a large amount of fish in the market caused a drop in price (the most notable example was 1903). Thus, the data show a correlation between catches, prices, and the intensity of the fisheries. They also allow us to draw conclusions about social factors influencing the size of the catches.

In the future we plan to study the connection between various parameters for smaller administrative regions, such as *uezd* and perhaps *volost* as well. We also intend to examine this connection over a longer period. It is quite possible that this would also help us to estimate the salmon population.

Fisheries and the Population Status of Salmon in the Twentieth Century

In the second half of the twentieth century significant changes occurred in the salmon population. It would appear that previously the high reproductive capacity of numerous and variable salmon populations had resulted in a large salmon population, while the local human population remained relatively small.

From the beginning of World War I to the mid-1920s, the salmon fisheries declined dramatically. This was caused by complicated and difficult political, social and economic circumstances during the war and the Russian Revolution. We have very little in the way of quantitative data for this period. Salmon fisheries had always required sophisticated and expensive fishing gear that varied according to seasonal conditions, and there was a dependence on the demand of the free market. It is therefore understandable that with the elimination of the free market this activity almost disappeared in regions where it was not essential for the local economy. This decline was further aggravated by a catastrophically low supply of materials for repairing fishing gear, salt and barrels for storage. These factors were further compounded by a sharp decrease in the number of fishermen.[39] It is possible that this downturn in the fisheries (and perhaps in timber-rafting as well) favoured the preservation and even an increase in the salmon populations by the mid-1920s, when the salmon fisheries began to recover.[40]

Up to the 1940s, the dynamics of the salmon catches were not much different from those observed at the beginning of the twentieth century, with the exception of the war period and the Russian Revolution. The population size of salmon is highly dependent on many natural factors, and until this period its dynamics reflected the natural cycle, which fluctuated between eight and eleven years. This cycle is determined mainly by global changes in temperature and other natural factors.[41] For example, research carried out as part of a large expedition in 1930-1933 that covered most of the salmon rivers of the Russian North

---

[39]Nikolsky, *Byt i promysly naseleniia*, 107-108; RGAE, f. 478, op.7, d. 3506, 3-53; and V.K Soldatov, "Ryby reki Pechory," *Trudy Severnoi nauchno - promyslovoi ekspeditsii*, XVII (1924).

[40]I. Yerofeichev, *Promysly Arkhangel'skoi gubernii* (Arkhangelsk, 1925), 18; and G.N. Glebov. "Puti razvitiia rybnogo promysla na Belom more v KFSSR," *Trudy pervoi nauchno - tekhnicheskoi konferentsii po rybnoi promyshlennosti KFSSR, 6* (Petrozavodsk, 1947), 313.

[41]Kazakov (ed.), *Atlanticheskii losos'*.

concluded that a sharp decline in the size of salmon catches, first observed in 1928, was due to natural fluctuations in the salmon population and not to over-fishing, as had been suggested initially.[42]

Figure 6
Total Catch Sizes of Atlantic Salmon in the Russian North

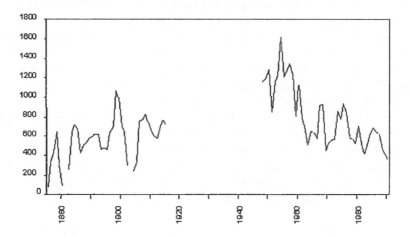

Source: Data compiled by Arkhangelsk Statistical Committee; and R.V. Kazakov (ed.), *Atlanticheskii losos'* (St. Petersburg, 1998).

In some regions a significant decline in fishing effort, which followed a drop in the price of salmon, was also observed. On the whole, during the 1920s and early 1930s the purchasing power of the Russian population was very low, and there was no longer the kind of demand for an expensive product like salmon, as in pre-Revolutionary Russia. Along with the decline in fishing effort, the low market demand for salmon meant that an increasingly high share of the catches went for local consumption. As a result, these catches were not officially recorded: according to data collected by the expedition mentioned above, this could reach fifty percent. In recognition of this phenomenon, L.S. Berg suggested adding twenty to twenty-five percent to the official catch figures in the region to get a more realistic estimation of actual catches.[43] Taken together with the fact that the collection of statistical data during this period had many shortcomings, this

---

[42]Berg, "Materialy po biologii siomgi."

[43]*Ibid.*

means that we will need to undertake additional research to uncover any data that exists on catches between 1920 and 1950.

Starting from the mid-1930s, catches began to increase rapidly, and this high level continued until the end of the 1950s, except during World War II (see figure 6). The organised catches were accompanied by an increase in poaching, which was connected with population growth in the North and the looming danger of over-fishing, and was accompanied by attempts to control the size of catches. In the beginning of the 1960s bans were put into place nearly everywhere. For example, at the Pechora River, by the 1960s the fisheries were concentrated at the mouth, and all fisheries further upstream or in its numerous tributaries were prohibited. This suggests that the main factor in the decrease of the salmon population was over-fishing. According to some recent estimations, illegal catches may still comprise up to one-half of total catches.[44]

By the middle of the twentieth century new factors had a negative impact on the salmon stock. The most important were timber-rafting, which is a source of pollutants in the spawning grounds and often blocks the home rivers; discharges of toxic substances into the rivers; and hydro-electric power plant dams, which prevent salmon from getting to the spawning grounds. The development of international fisheries in the Barents, Norwegian and other seas also exacerbated the decrease of salmon catches in the Russian North.[45]

A further decline in the catches was observed in the early 1980s, after which catches remained relatively stable during the late 1980s until a sharp fall occurred in 1992. Since then catches have not reached their previous levels, despite the fact that some of the factors that contributed to the decline in populations, such as timber-rafting, were ended in the mid-1980s. Currently, the available information allows us to estimate the population status of the salmon in about 130 rivers.[46] The results of this exercise show that in eighteen of these rivers the local populations of salmon are already extinct; in twenty-seven others the status of the salmon population is critical; and in thirty-six rivers the salmon

[44]V.G. Martynov and A.V. Zakharov, "Otsenka nelegal'nogo vylova atlanticheskogo lososia v reke Pechore putem interviuirovaniia" in A.I. Taskaev, *et al.* (eds.), *Simposium po atlanticheskomu lososiu: Tezisy dokladov* (Syktyvkar, 1990), 28; and M.Y. Alekseev, V.N. Pavlov and N.V. Ilmast, "Population Dynamics of the Atlantic Salmon (*Salmo salar L.*) in Some Rivers of the Kola Peninsula with Commercial Fisheries," in S.P. Kitaev (ed.) *Problemy Lososevykh na Evropeiskom Severe* (Petrozavodsk, 1998), 12-17.

[45]E.L. Bakshtansky, "Razvitie morskogo promysla atlanticheskogo lososia," *Trudy VNIRO*, LXXIV (1970), 150-176.

[46]Dmitry Lajus, Sergey Titov and Vasilii Spiridonov, "Russia," in *The Status of Wild Atlantic Salmon: River-by-River Assessment*, WWF Report (2001), 135-141.

populations are endangered. Hence, in fifty-two percent of the rivers the status of salmon populations is at various stages of suppression. In forty-two rivers the populations are under strain, and only seven to nine populations can be considered relatively safe. The best situation is at the Kola Peninsula, where the catches have been stable over the past several decades, although a decrease in the average size of the salmon has been observed.[47]

Comparison of Catches in the Onega River in the Seventeenth Century and Nineteenth-Twentieth Centuries

We used as a model the Onega River, because at this stage of our work we have a better set of historical records for this locality. The main difficulty in comparing catches in this river over a long period of time is that the sources are divided into different administrative regions. We have separate sets of data on the catches from the weir in Podporozhie, for the whole *volost* of Podporozhie (including catches at the weirs and the drift nets), for the whole of the river (including data not only from the *volost* of Podporozhie but for a few others upstream as well), and also from the town of Onega (sometimes they are given separately). The data on catches in the district of Onega include information on the catches from the river, from the marine fishing grounds near its mouth, and from the marine and river fishing grounds along the eastern coast of Onega Bay. It is therefore not always possible to find data for various periods that refer to the same area. Nevertheless, there are common patterns and relationships between some of the data sets. For example, between 1685 and 1691 the correlation coefficient for catches in three areas (Porozhskaya *volost* by peasants, Porozhskaya *volost* by a weir and Podporozhskaya *volost* by peasants) averaged 0.84. This makes it possible to extrapolate catch estimates in different areas with confidence. Nevertheless, we used extrapolation only within narrow limits.

The available data show a very high variability in catches during this period. For example, in 1615 at the Onega River 1760 salmon were caught, while in 1703 only seventy-three were landed, despite no obvious changes in the organisation of the fisheries. The observed variability is relatively smooth, i.e., catches in the neighbouring years are close to each other. The same holds true for the late nineteenth and early twentieth centuries.

As to the general dynamics of the catches, the maximum number for the Onega River was between four and five thousand fish in 1847-1851, 1908 and 1934-1937 (see figure 7). Catches in the range of 1500-2000, which were lower but still relatively high compared to those from neighbouring periods, were

---

[47]R.V. Kazakov and A.E. Veselov, "Populiatsionnyi fond atlanticheskogo lososia Rossii," Kazakov (ed.) *Atlanticheskii losos'*, 383-395.

observed in 1615 and 1689. Very poor catches (fewer than 100 fish) were recorded in the early eighteenth century, 1878, and in the 1980s.

Figure 7
Catch Sizes of Salmon (Number of Specimens) in Onega River

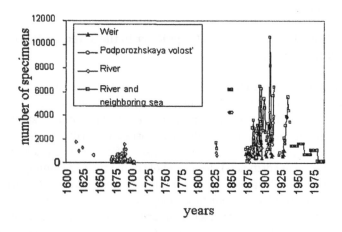

years

Source: See text.

For several centuries up to the 1930s, the main fishing gear at the Onega River was the weir, which probably evolved little over this period. According to a description by A.A. Krysanov, it was comprised of stakes, branches and twigs.[48] More than 500 men took part in its construction, which took about one month or 1000 man-days. In the seventeenth century, the weir belonged to the Krestny Onezhsky monastery, which evidently provided the materials for its construction. But a further search of the archives should be conducted to reveal the details of design and construction of the weir in earlier periods.

It is evident that in the twentieth century over-fishing and other man-induced factors, which had negative impacts on the salmon populations, began to appear. Their impact became especially obvious from the 1930s, when timber-rafting at the Onega River was quite intensive. The drop in catches was particularly noticeable from the middle of the twentieth century, and current catches are negligible. At the same time, there are no valid reasons to assume that over-fishing alone could have had a significant impact on the salmon's population size in the seventeenth century. Thus far we believe that fluctuations in catch sizes

---

[48]A.A. Krysanov, *Pomorskie promysly (Onezhskii uezd 1861-1916 gg.)* (Onega, 2000), 18.

in that period were caused by natural phenomena, particularly low average annual temperatures.

We have to add that the relationship between salmon and other marine species have not yet been investigated in detail. Post-smolt salmon are prey for a number of species, such as cod, birds, seals and whales.[49] It is not improbable that the high numbers of whales in the North Atlantic, together with the decrease of average temperatures in the seventeenth century, could also have had a negative impact on the size of the salmon populations.

Therefore, we conclude that the size of the salmon population in the seventeenth century was close to, or slightly lower than, that of the late nineteenth and early twentieth century. This conclusion is in conflict with the widely-accepted view about the impoverishment of the fish resources in the Russian North at the beginning of the twentieth century compared to its richness in the past. For example, Kuznetsov mentioned (without providing an exact reference) that in the vicinity of the town of Kola catches annually amounted to 393 metric tones, i.e., nearly ten times higher than at the beginning of the twentieth century.[50]

Our data on the catches at the Onega River are based on reliable sources, yet they contradict the notion of a sharp decrease in salmon numbers over the last few centuries. It can be added that data on catches in other rivers also support our conclusion. For example, at the Varzuga River the size of the catches in 1615 was about fifty-nine tons (32,820 fish), which is close to the average catch in the twentieth century. Therefore, the published data on the exceedingly high catches in the past should be verified carefully.

Some data allow us to estimate the average weight of salmon in the seventeenth century. For example, in the Varzuga River the average weight of fish collected for the tithes was 1.8 kg., which is smaller than in our times (between 1949 and 1951 it was about 2.6 kg.).[51] At the same time, at the Onega River at the beginning of the twentieth century the average weight was about 6.7 kg., whereas in the seventeenth century it was about eight kg., and for one of the documented years (1642) it was eleven kg. On the other hand, it is important to remember that as a rule tax collectors selected the largest fish for the tithes.

---

[49]J.R.G. Hislop and R.G.J. Shelton, "Marine Predators and Prey of Atlantic Salmon," Mills (ed.), *Salmon in the Sea*, 104-118.

[50]Kuznetsov, *Beloe more i biologicheskie osobennosti*.

[51]V.S. Mikhin, "Promysel semgi v reke Varzuge," *Izvestiia VNIORKh*, XLVIII (1959), 7-10.

Conclusion

We have thus far found a number of reliable sources for the estimation of the dynamics of the salmon populations in the basins of the Barents and White seas since the seventeenth century. While this essay is only a preliminary report, in our continuing studies we will compile data for periods not yet covered. We also intend to include more of the available indirect data in our analysis. We hope to obtain evidence which will allow us to describe the dynamics of salmon populations in a broad geographic area and separately for the main rivers and river basins. This in turn would allow us to evaluate changes in the population of the species. On this basis we will be able to analyse the role of human impact and climatic fluctuation in long-term dynamics.

We believe that the success of historical and ecological research on salmon, as well as on other widely distributed species such as cod, depends much on future international collaboration. It is quite evident that it is only by combining the data for different fisheries that we can understand the general dynamics of the fish populations, and the HMAP programme is designed precisely to facilitate such collaboration. International cooperation is also required to better understand the historical processes in one ecosystem – the Barents Sea area is a good example, since it contains an international fishery. We hope that this paper, however preliminary, will contribute to the general goal of HMAP collaboration.

The Danish Fisheries, c. 1450-1800: Medieval and Early Modern Sources and Their Potential for Marine Environmental History

Poul Holm and Maibritt Bager

Abstract

Primary source materials held in Danish archives are of great value for environmental historians seeking evidence from which to construct long time series relating to the inshore fisheries of regions bordering the North and Baltic seas. This paper focuses on documentary sources pertaining to these fisheries in the sixteenth and seventeenth centuries, but attention is also afforded to ways of extending the time series. Estate records reveal the diversity of Denmark's fishing interests when the catching effort reached a relative peak in the late Middle Ages (c. 1450-1590). At this time, haddock, rather than cod, formed the bulk of the North Sea catch, demonstrating that this species lived much farther south in the North Sea than it does today and suggesting that it was more abundant than it is today. The herring catch records reveal decadal to centennial shifts in abundance between the Limfjord and Bohuslen, possibly related to climate forcing. The historical evidence also reveals that Baltic cod seems to be related to a periodical pattern with high abundance in the seventeenth and late twentieth century interspaced with a period of low abundance. The paper examines why the Danish fisheries declined rapidly in the seventeenth century, with natural causes – climatic change, salinity changes and species competition – set against human social and economic factors, such as changing dietary preferences and the competitive forces of the market.

Danish Historical Sources

Before 1658, the Kingdom of Denmark comprised the provinces of present-day southern Sweden (namely Scania, Halland and Blekinge), as well as its current territories of Jutland, Funen, Sealand and Bornholm (see figure 1). The archival sources generated in these provinces were therefore assembled according to Danish administrative practice. Moreover, Norway, which is not covered in this paper, was under Danish control until 1814, and accordingly its records were

Poul Holm and Maibritt Bager

devised in line with Danish administrative practice. There is thus much uniformity in the information that relates to the highly saline-to-brackish waters bordering the North Sea, and the data concerning the Baltic, including the Skagerrak and Kattegat, and the brackish Limfjord. While a broad range of sources yields information on Danish medieval and early modern fisheries, the regional estate accounts of fiscal revenue are of special importance for the construction of long time series. These relate mainly to the fishing effort, for the taxes were levied either on the boat (the North Sea sand toll) or on the individual fisherman (the Baltic oar toll). The accounts of the King's catches on specific days at the Limfjord are another important source of evidence. Trade records, mainly port records and the Sound Toll Accounts, offer export measures, but also provide proxies for total catches. For the eighteenth and nineteenth centuries, a series of national surveys conducted by the Danish state offer a good deal of evidence on fish catches and the fishing effort. The potential utility of this array of documentary sources will now be discussed on a region-by-region basis in order to establish a baseline of seventeenth century information, which is then compared with evidence for the eighteenth and nineteenth centuries.[1]

Figure 1
Map of Denmark

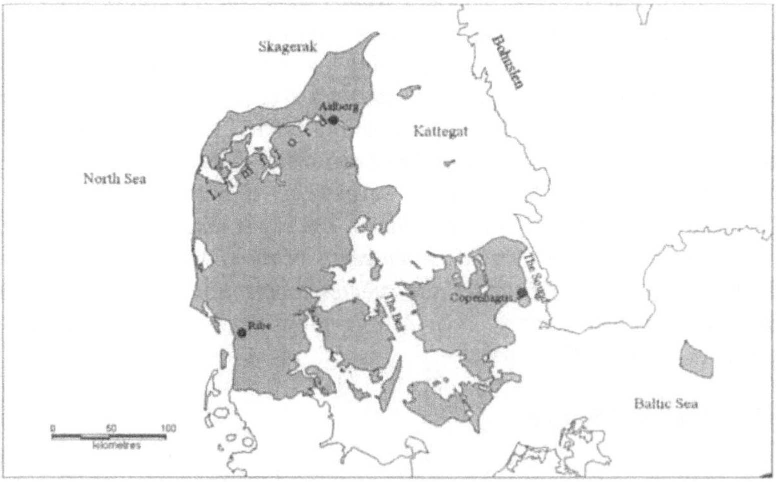

<hr />

[1]For an assessment of the economic importance of marine resources, see Poul Holm, "Catches and Manpower in the Danish Fisheries, c. 1200-1995," in Poul Holm, David J. Starkey and Jón Th. Thór (eds.), *North Atlantic Fisheries, 1100-1976. National Perspectives on a Common Resource* (Esbjerg, 1996) 177-206. Other valuable sources, such as the records of exports to the various Baltic and North Sea towns, will not be considered here because they constitute a vast remit which call for specialist evaluation.

The North Sea

The fisheries of the Danish North Sea coast were dispersed along the entire littoral, with the main concentration occurring around the Reef of the Horn, where haddock stocks were exploited. Taxes were charged on the boat, 1200 haddock being levied against large boats with crews of twelve men, while smaller boats paid half. The fishery was first mentioned in the thirteenth century and seems to have culminated in the first half of the sixteenth century when quantitative sources become available (see figure 2). In the second half of the sixteenth century, a long-term contraction set in. The decline in fiscal returns was offset by a new tax on the plaice fishery, which was apparently introduced in about 1550, and which for a time made up for the loss to the King in financial terms. By 1630, the fisheries were much reduced and remained very depressed through the following two centuries.

Figure 2

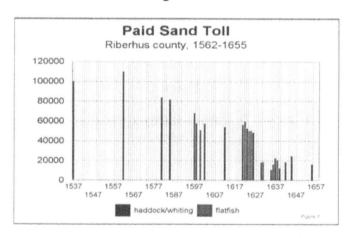

The tax returns indicate that a variety of species was taken (see figure 3). Evidently, the main fishery in the medieval period was for haddock, but flatfish were just as important by the late sixteenth century. The relative significance of these two categories of fish alternated again in the 1630s and 1640s when the haddock/whiting were caught in greater quantities, but altogether on a much reduced level. Other species were discernible later in the period, though it is highly probable that they were also taken in earlier times without being taxed. It was only when the main fisheries declined that they were brought within the taxation regime. The long-term trend for all species, however, was of almost uninterrupted decline. One point of dramatic decline stands out: the years 1627 and 1628, when Jutland was occupied by a Swedish force. The fishing settlements seem to have suffered a severe blow from which they never recovered. The long-term decline was not determined by this single episode but had already commenced before the end of the sixteenth century.

Figure 3

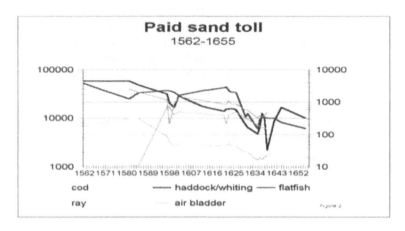

Figure 4

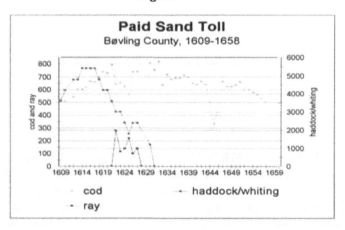

Farther to the north, in Bøvling County, the fishery was not as intense (see figure 4). In modern times, this part of the coast lies at the heart of a large inshore cod fishery, but its lack of intensity may be explained by the remoteness of the coast relative to the large markets of Hamburg and Schleswig. The tax returns indicate that the catch of both cod and haddock/whiting increased in the first decades of the seventeenth century. While the cod fishery was conducted from the northern parishes, the haddock/whiting fishery was only carried out in boats based in the southernmost parishes; indeed, it may have been located in the Reef of the Horn area as was the Riberhus fishery. By the 1620s, this fishery declined rapidly and was reported for the last time in 1630. The decline occurred at the same time as a new ray fishery developed, but this had disappeared before the end of the haddock fishery. The cod fishery was related to the ray fishery, as the accounts sometimes give a combined figure for the total catch of cod and ray. Moreover, the cod fishery declined somewhat during the period of the ray fishery. While the cod fishery peaked in the 1630s, the fisheries were stable through the 1640s if we disregard the figures for 1644-1645, which were influenced by another Swedish invasion. In the 1650s, the fishing effort declined. A couple of Jutland estates remain to be analysed, but it is known that by the mid-sixteenth century the total revenue of the northernmost areas of the North Sea coast amounted to roughly half the amount taken in Riberhus County.

We have no evidence for the Skagerrak coast. Without question, however, the most important fishery was conducted from the town of Skagen, which was immune from taxation and hence our sources of information are meagre. We do know that in 1583, Skagen had an armed citizenry of 335 men, and based on this number a guess of 400-500 active fishermen may not be wide of the mark.[2] In comparison, Riberhus County had a fishing population of around 1200 men.[3] Skagen's fish resources were, however, more plentiful because of the rich fishing banks around the Scaw.

The overall picture of the North Sea and Skagerrak fisheries before 1600 is one of general prosperity, with the maximum fishing effort occurring around the Reef of the Horn, with smaller centres around Thy and Skagen. The evidence does not at this point permit an assessment of overall fishing effort and catches, but further research may provide the necessary information. The evidence does bear out that the species composition in the southern North Sea was different from the twentieth century, with a large stock of haddock and ray.

The post-1660 estate records for West Jutland have yet to be investigated. Trade figures do, however, provide a clear indication of developments through the seventeenth and eighteenth centuries (see figure 5). The export of Ribe declined from one million flatfish in 1600 to 300,000 pieces in 1700, and halved over the next thirty years. The trade of the other main outlets, Hjerting and Ringkøbing, was just as bad, though the figures are less than complete due to the destruction of many customs records. The trade statistics for Skagen have

---

[2]C. Klitgaard, *Skagen Bys Historie* (Skagen, 1928), 36.

[3]Calculated from the total of 151 boats in the estate account for 1581 at an average of eight men per boat. The average is indicated by sources cited in J. Kinch, *Ribe Bys Historie* (2 vols., Copenhagen, 1869-1884), II, 863-864.

yet to be assembled. The figures leave no doubt, however, that West Jutland fish exports declined rapidly and that exports had almost completely ceased by the end of the eighteenth century. We shall return to the question of how to account for this change from a state of prosperity to poverty.

Figure 5

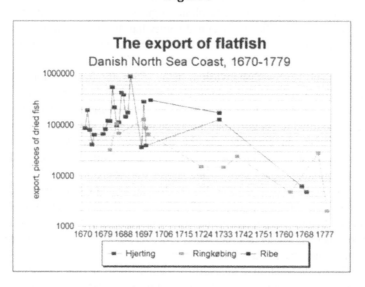

The Limfjord

The Limfjord ecosystem has been exposed to recurrent shocks during the last millennium which caused the water to change from highly saline to brackish in character. Archaeofaunal investigations have documented the fjord fishery of the Iron and Viking ages. It was dominated by plaice and eel at a time when the fjord must have been highly saline as it was open both to the north and west to North Sea water of 3.5% salinity.[4] Rises in the land level closed the entries to the fjord sometime in the eleventh century, preventing the inflow of salt water from the North Sea. By 1200, the fjord was reported to have a rich herring fishery. This presumably reflected the ability of the herring to adapt to alternating brackish and saline water conditions, as the fjord was its main feeding area through the spring before it migrated into the three-percent saline conditions of the Kattegat for spawning. In 1624, the fjord was opened to the west by a storm. The herring

---

[4]Inge Bødker Enghoff, "Fishing in the Baltic Region from the 5th century BC to the 16th Century AD: Evidence from Fish Bones," *Archaeofauna*, VIII (1999), 41-85.

stock seems to have suffered from the influx of North Sea water, as the fisheries were reported to be much depressed by the early 1630s. Over the next few years, the breach in the sand dunes closed and the stock recovered. A well-documented repetition of the scenario occurred in 1825, when a breach of the coastline caused the herring to be first caught in large numbers and then to become extinct within a decade. The fjord was now kept open artificially, and again occupied by plaice and eel.

Figure 6

Royal herring privilege
Liim Fiord 'kongekøb' 1600-1655

Figure 7

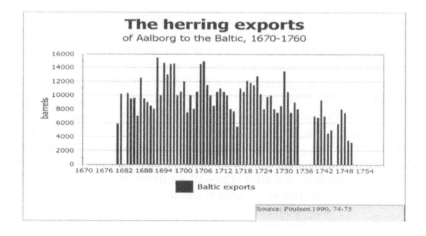

The herring exports of Aalborg to the Baltic, 1670-1760

Source: Poulsen 1990, 74-75

The herring fishery was mainly prosecuted by the use of large and very expensive pound nets owned by Aalborg merchants. The sixteenth century seems to have marked a peak in the herring fishery and witnessed an intense effort by the King to control fishing effort by strict regulations of mesh size and the allotment of fishing grounds to a limited number of pound nets. Evidence gleaned from the royal privileged purchase, the *kongekøb*, for the sixteenth and seventeenth centuries, together with the figures of Aalborg's exports to the Baltic for the eighteenth century, give an impression of the periodicity of the fishery (figures 6 and 7). The fishery seems to have been fully developed by 1518, when the *kongekøb* amounted to 528 barrels of herring.[5] In the seventeenth century, the annual *kongekøb* oscillated between 400 and 900 barrels in periods of good harvests and much less in bad times. Fishing was depressed in the early 1600s, and then peaked between 1610 and 1620. An evidential lacuna in the 1620s prevents us from following the impact of the 1624 flood, but the figures from 1631 reveal a fishery which was moving towards a new peak in the 1640s. Contraction had set in again by 1654, and there ensued a long depression, which is corroborated in other sources. Unfortunately, we are so far unable to link the *kongekøb* series with the trade statistics of the herring trade, which only begin in the 1670s. We do not therefore know if the remarkable catches around 1700 had been matched in earlier years. The trade statistics document a series of very good seasons, culminating in the peaks of the 1690s and 1706. From then on, the fishery followed a decadal pattern of ups and downs at a relatively high but declining scale until the early 1750s, when Aalborg's Baltic exports virtually ceased.

Aalborg also conducted an export trade to Norway, for which we only have information dating from 1722 to 1764 (figure 8). This additional data modify the impression of the fishery gained from the Baltic trade figures. On the one hand, the basic picture of the rise and decline of a significant fishery is confirmed. In addition to the Norwegian export we must allow for an internal Danish trade, which would certainly have brought total trade well above 50,000 barrels of herring in the 1720s. After this peak the Norway trade declined through the next three decades to a very low level in the 1760s. On the other hand, the Norway trade figures show that the fishery, although much reduced by 1760, did continue. While the overall contraction may be explained by a declining stock, the complete cessation of the Baltic trade after the mid-1750s must be related to another phenomenon, namely the growth of the Swedish herring fishery. From 1756, the herring fishery in the Skagerrak and Kattegat archipelago of Bohuslen began a half century of unrivalled dominance of the northern European herring market. The Aalborg merchants, already hit by the decline of their own herring stock, were unable to compete with the sudden and

---

[5]Jan Kock, "Næringsliv og samfærdsel," *Aalborgs historie*, I (1992), 355.

massive availability of cheap Bohuslen herring. Aalborg's herring fishery was thus much reduced, and while it continued to compete favourably in Norway, where Danish merchants enjoyed customs protection, the town lost the Baltic market.

Limfjord herring began to be caught in large quantities after 1800, when the Bohuslen fishery suddenly collapsed. The period to 1832 saw a renewed rise of the Limfjord herring fishery. The towns of Aalborg and Nibe generated tremendous wealth at a time when the rest of the country suffered the consequences of Denmark's misfortune in the Napoleonic wars. Extant records of the catches per day through one year of a typical fishing weir reveal very high productivity. In 1825, however, the sand dunes which blocked the inflow of salt water from the North Sea to the Limfjord were breached by a flood, and the fjord turned salty after 800 years of brackish conditions. The consequences to the herring stock seemed beneficial at first. Catches rose to an all-time high over the next three years, but then dwindled dramatically. By 1832, the fishery was considered extinct.

Figure 8

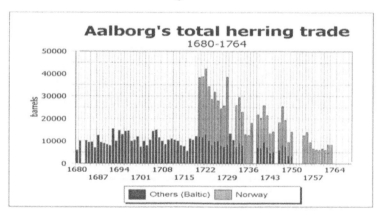

The Bohuslen Fishery

The Aalborg and Bohuslen fish merchants seem to have experienced alternating fortunes from the sixteenth to the early nineteenth century. Aalborg's bleak 1660s were good years for Bohuslen, while the decline of Aalborg in the 1750s was matched by an unprecedented growth of the Bohuslen fishery.[6] Data

---

[6]For this paper, which is concerned only with the Danish sources, we restrict ourselves to the sixteenth-century evidence. Bohuslen was then a Norwegian province and therefore under Danish administration.

pertaining to the fishery are available annually from 1557 in the Sound Toll
Tables, which comprise registers of ships entering the Danish Sound and paying
dues according to the nationality of the owner of goods on board. As Danish,
Norwegian, Lübeck and Swedish ships were exempt from the tax, the records are
less than perfect. But they do provide minimum figures of the transit trade and,
if treated cautiously, can provide us with important information.

Figure 9

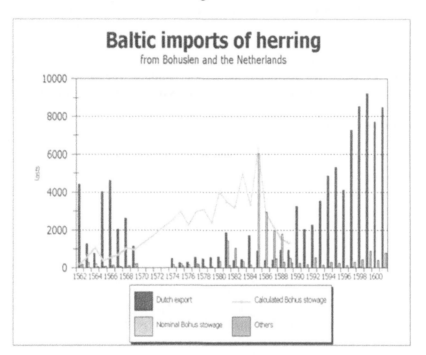

Figure 9 summarizes the information on ships carrying herring through
the Sound. Dutch exports to the Baltic clearly dominated the picture in the 1560s,
but when the Dutch Revolt broke out, they virtually terminated for twenty years.
This coincided with the rise of the Bohuslen fishery, which is reported to have
taken place in 1556. Most of the Bohuslen trade was carried out by Danish and
Norwegian ships, so the cargoes went untaxed and therefore mostly unregistered
through the Sound, although the very fact of the passage of a toll-exempt ship
was very often registered. Consequently, the nominal stowage of herring from
Bohuslen shows as a very modest row of columns in the graph. In 1585,
however, the customs officers seem to have been instructed to make full
inventories of all toll-exempt ships. For a few years, we therefore get a much

fuller picture of the cargoes passing through the Sound. This information can be used to make a retrospective calculation of the stowage of herring aboard the ships that had been registered as passing. The estimated stowage of herring from Bohuslen shows rapid growth in the early 1560s to a level of 1000 lasts, or 12,000 barrels of herring. In the 1570s, when the Dutch virtually ceased their exports, the Bohuslen trade quickly built up to a level of around 50,000 barrels, peaking at 60,000 barrels in 1586. The trade then rapidly declined and had virtually ceased by 1589, although a couple of years in the 1590s witnessed some herring coming from Bohuslen. The Dutch had a superior product and the Bohuslen fish merchants were no match for them.

The appearance and then disappearance of the Bohuslen herring was probably caused by the natural periodicity of the stock, which seems to have been linked to the North Atlantic Oscillation.[7] Yet the significance of the withdrawal of the Dutch from the market, and then their re-entry in the 1590s, cannot be neglected in any assessment of the scale and duration of the Bohuslen fishery. While the business started slowly and conceivably reflected increased abundance, the collapse of the fishery and the lingering presence of catches in the 1590s indicates the force of sudden changes in market conditions. As with Aalborg's trade with the Baltic and Norway, the Bohuslen fishery may have continued to a degree in spite of the loss of the Baltic market, as reflected in the Sound Toll Records. A combination of natural and economic factors seems to explain the specific development of the sixteenth-century Bohuslen herring period.

The Sound Fishery

Another factor in the development of the Bohuslen fishery was the contraction of Denmark's most important fishery, the Sound fishery for herring, in the second half of the sixteenth century. The exact timing of the decline is not known, but signs of a crisis were evident in 1547, though an estate meeting of 1558 continued to stipulate that free access to the herring market of the Sound was the first privilege of a nobleman. By the 1580s, the herring town of Stege was depressed due to the crisis in its staple trade, while in the early seventeenth century the King doubted if German merchants were still making their customary annual appearance in the herring market, which was necessary if they wanted to keep their privileges.

This was a dramatic downturn in a fishery that for a couple of centuries had provided Denmark's main export commodity. The trade was in the hands of German merchants, but the catching activity was mainly undertaken by Danes.

---

[7]Brian R. MacKenzie, *et al.*, "Ecological Hypotheses for a Historical Reconstruction of Upper Trophic Level Biomass in the Baltic Sea," *Canadian Journal of Fisheries and Aquatic Sciences* (under consideration).

Unfortunately, the estate records cannot be used to reveal the size of the fishery, since German traders were exempt from the jurisdiction of the Danish authorities. Evidence has been utilised by Christensen, however, to indicate that the fishery around 1400 yielded a total export of 200,000-300,000 barrels of herring. The catches were taken not only in the Sound proper but also as far into the Baltic as the island of Bornholm. These estimated catches can be related to modern records. A standard Rostock barrel of herring weighed 117 kilograms, including brine. As one barrel of salt was needed for three barrels of gutted herring, the net herring content was approximately 100 kilograms of wet fish.[8] The peak medieval export thus corresponded roughly to a catch of 30,000 tons. In addition, fishermen kept some fish for their own consumption and sale. This component of the catch perhaps added 100,000 barrels, so that the medieval Sound fishery produced about 40,000 tons of herring.[9] Around 1900, before the introduction of motor boats and the trawl, catches in the Danish part of the Sound were around 8000-9000 tons per year.[10] Catches on the Swedish side were probably larger, and we may therefore estimate a total catch in modern times around 20,000 tons. The medieval fishery was therefore twice as productive as its modern counterpart, though its execution was much more labour intensive, with drift nets suspended at night from thousands of small boats during two or three hectic months in the autumn. A rough calculation indicates that a total of 17,000 fishermen may have been employed in the fishery around 1400, with an additional 8000 engaged in related trades.

Catches varied greatly from year to year, the herring being virtually absent in years such as 1402, 1425, 1469, 1474 and 1475. In 1494, Danish herring exports (including the Limfjord) probably amounted to only 100,000 barrels, while the labour force had declined to around 6000 fishermen. In the 1520s, the Sound fishery experienced its last great period. According to the bailiff of Lübeck, dues were paid on no fewer than 7515 vessels seeking to enter

---

[8]*Kulturhistorisk Leksikon*, IV, 343-346; and *Diplomatarium Danicum*, IV, No. 1 (1984), no. 128.

[9]According to the *Modbog*, the judicial code for the Scanian fish market, no fisherman was allowed to salt more than six barrels of herring for his own use, with the remaining catch being sold fresh to the merchants. With an estimated total number of fishermen around 1400 of 17,000 men, we should add about 100,000 barrels for private consumption, making an estimated peak catch of 400,000 barrels.

[10]*Fiskeriberetning 1894-95, 1904-05* (Copenhagen, 1895 and 1905), Nordsjælland, Sydsjælland, Øresund, Lolland and Bornholm. In 1904, 64,451,600 herring were caught. If the average weight is estimated at 150 grams, this corresponded to 9668 tonnes. See L.G. Sjöstedt, *Barsebäcks fiskeläge* (Malmö, 1951), 87, for some practical experiments demonstrating the size and weight of medieval herring.

the fishery. With an average crew of five men, this meant that approximately 37,500 fishermen were involved in the business, each of whom was required to pay 240 herring to the King. Since a typical barrel contained 840 herring, and as the total revenue amounted to some 7703 barrels of herring, the total number of fishermen was somewhat lower than the estimate above, but still a staggering 26,960 men.[11] Unfortunately, the total catch for any one year in the 1520s is not known, but there can be no doubt that several hundred thousand barrels were produced annually in that decade. When a yearly production figure is available, the Sound herring fishery was already declining. This was in 1537, when the Danish customs officer at Falsterbo reported a catch of 96,000 barrels, while total Danish herring fisheries – from the other fishing ports on the Sound, Bornholm and the Limfjord – according to his and his colleagues' judgement amounted to around 360,000 barrels.[12] The Falsterbo figure compares favourably with the low catch of 60,000 barrels in 1494.[13]

The Sound fishery seems to have experienced two extraordinary peaks around 1400 and again in the 1520s. The value of the catches did not compare, however, because of the long-term price inflation of agricultural products as compared to marine products. From 1450 to 1550, the price of herring halved relative to grain.[14] Nevertheless, the value of the business encouraged the Danish nobility to invest considerably in the fisheries during the first half of the sixteenth century.[15] They contributed capital and labour from their estates, and the Sound fishery attracted crews from all over south Sealand and even from Jutland.[16] In the herring period, the most active towns were reported to have been almost

---

[11]*Danske Magazin*, VI (1836).

[12]D. Schäfer, *Das Buch des lübeckischen Vogts auf Schonen* (Halle, 1887), V. See also Mikael Venge, *Fra åretold til toldetat. Dansk Toldhistorie I* (Viborg, 1987), 72-76.

[13]Schäfer, *Das Buch*, 126-127. On repression of the Germans, see Kr. Erslev and W. Mollerup, *Danske Kancelliregistranter 1535-1550* (Copenhagen, 1881-1882), 1536 Malmøs reces 6 april.

[14]W. Bauernfeind, *Materielle Grundstrukturen im Spätmittelalter und der Frühen Neuzeit: Preisentwicklung und Agrarkonjunktur am Nürnberger Getreidemarkt von 1339 bis 1670* (Neustadt, 1993).

[15]Erik Arup, *Danmarks Historie* (3 vols., Copenhagen, 1932), II, 417.

[16]Bjarne Stoklund, "Bonde og fisker. Lidt om det middelalderlige sildefiskeri og dets udøvere," *Handels- og Søfartsmuseets Årbog 1959*, 101-122.

deserted during the fishing season.[17] It is little wonder that the decline of the Sound fishery in the mid-sixteenth century caused many fishermen to migrate in search of new grounds, as happened when Elsinore fishermen regularly sailed to Bohuslen from around 1560. Others engaged in fisheries that had not been exploited hitherto.

Unfortunately, the estate records of the Sound area survive only from the last decades of the sixteenth century and have not yet been fully investigated. One estate account, Kronborg Len, was investigated by Søren Frandsen and Erik A. Jarrum (see figure 10). The tax was paid on the oar, and the returns calculated in terms of eighty herring (the *ol*), which probably was the amount paid per man. The graph therefore reflects the number of active fishermen per year. Activity was at a peak in 1585, and one is left wondering if this reflected fishermen participating in the Bohuslen fishery rather than the Sound fishery, which according to all the qualitative evidence was already depressed by this time. The number of fishermen halved between 1585 and 1620, when the qualitative evidence suggests that the Sound fishery enjoyed a brief recovery. Numbers reduced again to one-third over the next twenty years. After this dramatic decline, the number of fishermen stabilised, with some active years around 1700, and again in the 1740s, probably reflecting a periodic upsurge in the herring stock.

Figure 10

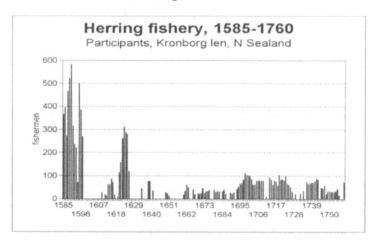

The tax returns of Malmøhus County, which have been preliminarily examined for this paper, do not include the fishing settlements along the coast of Skåne.

---

[17]*Ibid.*

The records, which are the household account of the vassals of Malmøhus castle, nevertheless give an insight into the diversity of the fishery. Tax was paid on herring, cod, garfish, large cod, ling, whiting, ray, eel, salmon, haddock and flounder on the lands belonging to the estate in the years 1565, 1572, 1575 and 1582. This diversity is not found in the seventeenth century. In the accounts of Malmøhus for 1624, 1625 and 1626, the list of taxable fish comprises just herring, cod and eel.

The estate records of the island of Møn also indicate that a dramatic decline occurred in the second half of the sixteenth century. In the 1540s, there were eleven fishing villages on Møn according to the tax lists, while the tax collectors only noted four in the 1620s, including one which had been deserted. Two early fragmentary lists point to the considerable tax revenue generated by herring catches, while later and more complete evidence from the early seventeenth century reveals a very limited revenue yield from the fishery. The herring fishery seems to have become very marginal, while the evidence indicates that a short-lived surge marked the cod fishery in the 1620s.[18]

The Cod Fisheries of Southern Denmark

The Nordic Seven Years' War (1563-1570) obliged the Danish Navy to solicit food from every corner of the Kingdom. White fish, especially cod, was eminently suited for naval purposes. As early as October 1563 the King endeavoured to buy salted cod for the navy from the islands of Anholt, Langeland, Møen, Bornholm and Kullen, and from the towns of Skagen and Simrishavn.[19] By year's end, the King had ordered his men in Jutland to buy cod and to ask for stock lists from all merchants to prevent them from exporting the fish.[20] Herring could be sold at good prices abroad and was therefore exempt from the export ban, while cod was mainly caught for domestic and north German markets and could therefore be reserved for the navy without heavy financial loss.[21] This urgent quest for cod generated a welter of documentary sources which reveal a previously unknown cod fishery. This may have been stimulated by the new demand, but the basic catch structure must have existed

---

[18]Jacob Svane Hansen, "Fiskeri på Møn i første halvdel af 1600-tallet. Belyst gennem Stegehus Lens regnskaber" (Unpublished BA thesis, Aarhus University, 2000).

[19]*Kanc. Brev. 1561-65,* 16 October 1563.

[20]*Ibid.*, 28 December 1563.

[21]*Ibid.*, 5 April, 9 August and 16 October 1563; and 14 January and 3 September 1564; *Regesta Diplomatica historiæ danicæ* II, No. 1, 20 March 1564, No. 4725; and *Kanc. Brev. 1561-65*, 3 September 1564.

before the war. Such is the case with the fishermen on the island of Møn by the Sound, for instance, who had long since been involved in the herring fishery. In early March 1564, they were catching cod in the Sound.[22] Archaeological evidence demonstrates that relatively new fishing settlements boomed as a result of naval demands, with long-lining spreading throughout the country.[23] A full investigation of the estate records will not only permit a more detailed picture but also offer a fuller insight into the interplay between cod and the herring fishery, which so far dominates our understanding of the development of Danish medieval and early modern fisheries.

The Bornholm Fishery

On the Baltic island of Bornholm, the annual tax lists of the crown's estate are preserved for nearly every year from 1597 until 1660, and in most years herring and cod fishermen are noted. From 1597 to 1612, the tax records are quite detailed, and it is possible to quantify the amount of taxable fish, as well as to establish the contribution made from the various sites along the Bornholm coastline. According to the accounts, every man had to pay 50-100 tax-herring (*told-sild*), in the form of salted herring in barrels, to participate in the herring fishery. An appendix to the tax list of 1607 states that 105 people paid tax-herring in that year. The cod fishery was likewise taxed on an individual basis, the tax-cod (*told-torsk*) being paid in the form of dried cod. After 1609, the payment of both herring and cod was made salted and barrelled, and it was no longer a tax by person but a constant duty on the rent of farms near the coast. But the fact that the peasants were relieved from paying the duty in poor years allows us to use it as a rough indicator of the fishing effort.

The crown's tax list from the seventeenth century gives an impression of the periodic nature of the fishery (figures 11 and 12). By 1700, the cod and herring fisheries were rich, but from the turn of the century both declined significantly. During a ten-year period, the tax-herring payment more than halved. Thus, in 1597, eighteen barrels of salted herring were paid by the island, while in 1607 the payment was only seven barrels. In the following years the herring fishery stabilised at an even lower level, until a new peak occurred in 1615. Then, no record of income from the fishery is found in the tax accounts until 1627, when Bornholm herring was again taxed. Down to 1644 the fishery appears to have been stable, but the records do not mention herring thereafter.

---

[22]*Kanc Brev 1561-65*, 13 March 1564.

[23]*Kulturhistorisk Leksikon, vide* Krogfiskeri; S. Frandsen, "De nordsjællandske fiskerlejer," *Bygd* (1987), 2-6; and H. Berg, L. Bender Jørgensen, and O. Mortensøn, *Sandhagen. Et langelandsk fiskerleje fra renaissancen* (Rudkøbing, 1981).

The cod fishery shows a periodicity on a somewhat longer wavelength. The payment of tax-cod indicates that a peak in the fishery occurred at the end of the sixteenth century. The fishery seems to have declined then, with cod missing from the tax lists between 1616 and 1627. In the early 1630s, the fishery entered another period of growth, building up to a new peak in 1655.

Figure 11

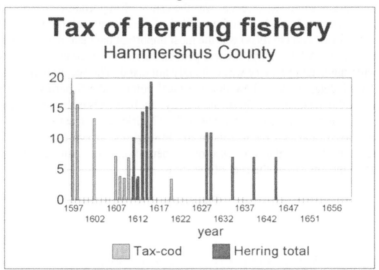

Figure 12

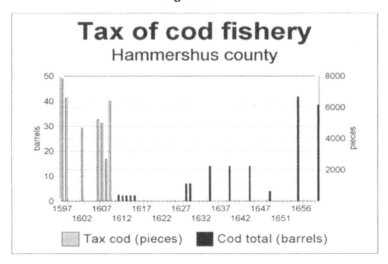

The Blekinge Fishery

The Blekinge fishery did not fluctuate in line with that of Bornholm. Rather, it was characterized by decline and stagnation throughout the first half of the seventeenth century. The crown's tax lists from Kristianopel County indicate a continued decline in the extraction of marine resources in the Blekinge fishery during the seventeenth century. This, the easternmost region of ancient Denmark and present-day southeast Sweden, was the centre of important cod and salmon fisheries (figures 13 and 14). In the tax lists of the crown preserved from 1604 until 1659, dried cod (*spidfisk*) was the most important payment made in respect of the fisheries. Additionally, tax was paid on barrels of salted cod, salmon and eel, while the herring fishery was evidently not taxed. The accounts from 1604 to 1613 are detailed, and show that the focal point of the fisheries was Øster herred, in the southeastern tip of the county. In 1609, Øster herred delivered seventy-six percent of the total revenue of 30,474 dried cod. The fishermen were not only taxed directly, but also indirectly through the crown rent of the farm (*jordebogsafgiften*), which was fixed over a period of years. After 1613, crown revenue came only from the rent of the farm.

Figure 13

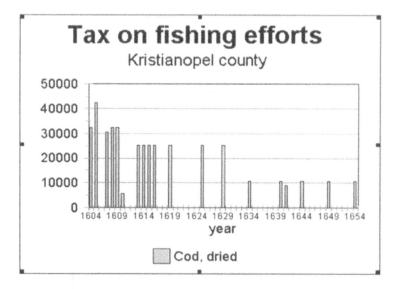

Figure 14

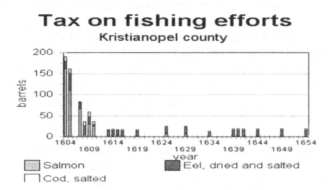

The duty on the cod fishery was reduced by seventy-five percent, the salmon fishery almost ceased to exist, and the eel fishery declined as well. The cod fishery reached its zenith in the beginning of the century, when 42,364 dried cod were paid to the king in 1608. This year was also a good year in the salmon fishery, with a total tax income of 110.5 barrels of salmon, though 1607 had been an even better year with a total revenue of 160 barrels of salmon. Yet in 1609 these high catches ceased, and the income from the salmon fishery disappeared from the accounts. The cod fishery was also in decline, but stabilised at a lower level in 1613, and until 1630 the annual duty amounted to approximately 25,000 dried cod. Samples of the accounts from the years 1634, 1639, 1640, 1641, 1644, 1649 and 1654 show that the cod fishery declined even further by 1634, leveling off at a duty of 10,600 dried cod. The decline in the cod fishery was reflected in the decline in the number of dried cod paid to the crown, but the revenue of barrels of salted cod was also reduced from eleven barrels in 1609 to eight barrels in 1610. By1613, the payment had reduced to three barrels annually. The payment of eel was not reduced by the same degree, yet the income from this fishery also diminished from the beginning of the seventeenth century to stabilise from 1613 at an annual income of thirteen to fifteen barrels.

The Decline of the Fisheries in the Eighteenth Century

The decline of the Danish fisheries in the late seventeenth and eighteenth centuries was evident to contemporaries. By the last quarter of the eighteenth century, Danish writers much deplored the state of the Danish herring fisheries. They could only watch with envy as the herring fisheries of Bohuslen experienced an unprecedented boom, with Aalborg's fishery once again rising in competition. The other Danish fisheries of the period, however, hardly receive mention and seem to have been ignored by most commentators due to their impoverished state. The once thriving fishing town of Skagen reached its low water mark, with a much reduced population living in stark poverty. The West Jutland fisheries exported little to Hamburg and north Germany, the catches being mainly bartered for agricultural produce with the hinterland.[24] The fisheries in North Zealand were not much better; the number of boats in Gilleleje declined from eighteen to seven between 1760 and 1785, while the fishermen complained bitterly about Swedish competition when the Bohuslen fisheries began.

As the fisheries contracted, fishermen drifted to other trades, especially the merchant shipping industry and the navy. While tens of thousands were involved in the late medieval fisheries, a mere 5000 people were so engaged in 1770, even though the total population had increased by as much as fifty percent. The second half of the eighteenth century was a time when merchant shipping flourished, and new jobs opened up for able-bodied seamen in world- wide trades.[25] Commentators remarked that while the fisheries provided the sailors for the growing merchant marine, the fisheries were a shrinking business. By all accounts, marine catches were at an historic low around 1800.

Causes of the Decline

There is no single explanation for the contraction of Danish fisheries in the late sixteenth and early seventeenth centuries. Rather, a combination of factors, both acute and chronic, environmental as well as economic, appears to have been at play. A distinction should also be made between cyclical variation and irreversible decline when discussing the long-term trends in the fisheries.

Economic Factors

Prices

The obvious backdrop to declining effort is dwindling returns on labour input. Unfortunately, there are no relevant price series for Danish fish. There is evidence to indicate that after a peak around 1450, fish prices fell throughout the sixteenth century relative to agricultural prices. Arnved Nedkvitne has published three series of data pertaining to the purchasing power of dried cod to grain on

---

[24]Poul Holm, *Hjerting. En maritim landsby midt i verden, 1550-1930* (Esbjerg, 1992).

[25]Poul Holm, "Kystens erhverv og bebyggelse, 1500-2000. Bidrag til Kulturhistorisk bygdeinddeling af Danmark," in Per Grau Møller, Poul Holm and Linda Rasmussen (eds.), *Aktører i landskabet* (Odense, 2000).

the Dutch, English and Norwegian markets, which throw light on the decline.[26] Whereas one kg. of dried fish would have bought fourteen kgs. of wheat on the London market around 1400, at the end of the sixteenth century it bought only six kgs. Similar evidence from Holland shows that dried cod lost almost half its purchasing power relative to rye during the sixteenth century, and on the Bergen market the purchasing power more than halved between 1400 and 1500, and halved again over the next hundred years. A similar development concerning herring can be calculated from the German evidence presented by Bauernfeind; between 1450 and 1550 herring lost half of its value relative to rye in the Nuremburg market, and continued declining in the seventeenth century, except for some very good years around 1617-1622.[27] Evidence from the neighbouring north European countries concerning both cod and herring thus shows a significant drop in the price of fish relative to agricultural products from the late medieval to the early modern ages. In the Netherlands, the relative price fall was counterbalanced by the growth in deep-sea catches by larger and more productive ships. In Denmark, fishermen were apparently unable to afford larger ships and were pushed out of the fishing sector by poor prices.

Table 1
Prices and Price Relations, Herring and Bread, Copenhagen, 1721-1800
(Skilling per *skippund)*

	Bread	Autumn Herring	Fish/Bread Ratio	Spring Herring	Fish/Bread Ratio
1721-1730	260	568	218	320	123
1731-1740	272	554	204	301	110
1741-1750	148	638	432	374	253
1751-1760	251	530	210	261	104
1761-1770	234	554	237	301	129
1771-1780	413	570	138	294	71
1781-1790	615	672	109	377	61
1791-1800	401	713	177	462	115

Source: Calculations based on tables in Astrid Friis and Kristof Glamann, *A History of Prices and Wages in Denmark, 1660-1800* (Copenhagen, 1958).

---

[26]Arnved Nedkvitne, *"Mens Bønderne seilte og Jægterne for."* Nordnorsk og vestnorsk kystøkonomi 1500-1730 (Oslo, 1988).

[27]Bauernfeind, *Materielle Grundstrukturen.*

The adverse trend in fish prices relative to agricultural prices continued through the eighteenth century. In table 1, the price series from the Copenhagen fish market shows that herring became cheaper as compared to the price of bread, thus worsening the purchasing power of fishermen relative to peasants. The spring herring came from the Limfjord, while the autumn herring most probably came from the Sound and North Zealand fisheries. The only notable exception to the price decline occurred in the 1740s when the fishermen must have experienced a time of rare and welcome prosperity. Reminiscences of the good times are to be found in the reports made to Chancellor Oeder some thirty years later, which generally lament the present and long for the good old days; they also specifically record data that seem to corroborate a decline within the past generation. Only in the last decade of the century did fishermen experience some growth in purchasing power relative to bread prices.

Capital

In the late sixteenth century, Danish noblemen reduced their maritime investments to concentrate on agriculture and stock raising.[28] As late as 1558, when the Båhus fisheries were growing, the right of the nobility to participate toll-free in any herring fishery throughout the country was promulgated by the estate assembly.[29] Later this freedom became of no importance, and the nobility began a political battle to restrict the freedom of their peasants to fish and trade; while the King occasionally supported the freedom of the peasants to leave their soil, the towns supported the noblemen. In South Zealand, the towns suffered because of the decline in the fish trade and attempted to stop peasants from sailing in order to keep the transport of goods for themselves. The combined efforts of nobles and towns succeeded by virtue of a series of royal commands.[30] The prohibitions meant, however, that the peasants lost interest in keeping a boat altogether; they not only stopped trading but also fishing. Before 1600, they had been effectively bound to the land, and a strictly agricultural system had developed. The late sixteenth century witnessed the development of domain manors, and this entailed the employment of a workforce which could not be absent for two or three months in the harvesting period. If, or when, the herring returned in great numbers, there was only a small group of professional

---

[28]Helge Gamrath and E. Ladewig Petersen, *Danmarks historie. Perioden 1559-1648* (Copenhagen, 1980), 409.

[29]*Corpus Constitutionum Daniæ*, I-VI (Copenhagen, 1887-1918), I.

[30]F. Martensen-Larsen, *Hav, fjord og handel. En studie i handelsveje i Nordjylland i tiden indtil 1850* (Herning, 1986) 151, note 10.

fishermen to catch it. The custom of recruiting crews from the surrounding agricultural areas had stopped, and economic interest had turned effectively away from the sea. One might expect that the large-scale medieval fishery would have left a lasting heritage in spite of the sixteenth-century contraction. But, on the contrary, southeast Denmark was described in the eighteenth century as a region where the peasant was tied to the land, despite the proximity of the sea. Net fishing may have demanded many hands, but only a few skilled fishermen. When money deserted the fishing industry, fishing was abandoned, while the professional fishermen congregated in a few fishing villages in the vicinity of big towns like Copenhagen, Malmø and Helsingør, where they could still find a local market for their products.[31]

Consumption

Many factors may explain the relative price decline. After the Protestant Reformation, "Catholic practices" such as the eating of fish at stipulated times were given up. When fasting regulations were abolished, the wealthy turned to a meat diet even on traditional "fish days." Northwest European consumption habits gave greater priority to meat and poultry; declining fish consumption was evident in Britain and the Scandinavian countries by the seventeenth century. Even so, fish remained in demand as a source of cheap protein for the labouring poor, especially the expanding numbers of seafarers who needed nutritious and well-preserved food for long voyages, but the demand for well-preserved, high-quality fish contracted.[32]

Practically no work has been done on Danish food consumption patterns in the sixteenth and seventeenth centuries, but it seems likely that the dietary changes largely occurred during the first half of the seventeenth century rather than immediately following the Reformation.[33] Fish consumption in rural areas probably changed more slowly than in the towns and among the nobility. By the mid-eighteenth century, the yearly ration of herring for a Zeeland agricultural

---

[31]Stoklund, "Bonde og fisker," 119.

[32]E.F. Heckscher, *Sveriges ekonomiska historia från Gustav Vasa* (2 vols., Stockholm, 1936-1949); C.L. Cutting, *Dishsaving. A History of Fish Processing from Ancient to Modern Times* (London, 1955); and F. Grøn, *Om Kostholdet i Norge fra omkring 1500-tallet og opp til vår tid* (Oslo, 1942).

[33]Lilli Friis, "Æde og drikke," in A. Steensberg (ed.), *Dagligliv i Danmark* (Copenhagen, 1969), 419-423. There is a rich source material in estate food lists.

worker was reckoned to be one quarter of a barrel, an intake that had not changed much since the Middle Ages.[34]

Competition

The fact that Denmark's fisheries crumbled while those of the Dutch expanded, was partly due to the fact that the Danes concentrated on inshore, lightly-salted products (and in the case of the Sound for the top end of the market), while the Dutch went deep-sea for large quantities of heavily-salted fish. The superiority of the Dutch enabled them to corner the rest of the market for high-quality herring. Dutch dominance led to the abandonment of the Danish-Norwegian coasts by thousands of people. After a brief pause during the Spanish War, when Bohuslen had its heyday, the Dutch returned around 1590 to supply the Baltic market. In the face of their plentiful, high-quality supplies, the Danish-Norwegian fishing industry succumbed. Dutch fisheries peaked in the first half of the seventeenth century, when 500 to 600 *buisen*, each of fifty to sixty tons burthen, worked by a total labour force of 7000 fishermen, brought home a total of 20,000 lasts, or 240,000 barrels, per year.[35] One of the reasons for the success of the Dutch was their much higher productivity. The average catch per fisherman was thirty-four barrels, compared with a catch per fisherman in the Sound fishery of around fifteen barrels. By 1590, the Dutch had regained control over the fishing grounds in the North Sea, and the clearances of Dutch herring through the Sound rose to an unprecedented height of almost 10,000 lasts by the end of the decade. This supply may have forced prices down and rendered the South Scandinavian fisheries unable to compete on the market. Vastly increased Dutch cod fisheries may also have adversely affected the south Scandinavian fisheries.

Environmental Factors

Short-term environmental changes undoubtedly played a decisive role in the depression of the Limfjord fishery around 1630, and again in the irreversible contraction that set in around 1830. The barrier between the fjord and the North Sea in the west was breached by floods in 1624 and 1825. While fish catches certainly increased in the years immediately after the second breach, the same

---

[34]Heckscher, *Sveriges ekonomiska historia*, xxv (note referring to 285).

[35]See Jaap Bruijn, "Dutch Fisheries: An Historiographical and Thematic Overview," in Holm, Starkey and Thór (eds.), *North Atlantic Fisheries*, 112-113. See also Jan de Vries and Ad van der Woude, *Nederland 1500-1815. De eerste ronde van moderne economische groei* (Amsterdam, 1995).

may also have happened in the 1620s. But by 1630 and 1832, respectively, the fisheries were reported to be much depressed. The herring, which was accustomed to brackish water, was probably severely reduced by the intrusion of salt water. In the 1630s, the dune barrier was built up by the natural sand drift of the North Sea coast, and the herring stock recovered.[36] The process would perhaps have repeated itself in the nineteenth century had it not been for dredging and coastal works that were implemented to keep the fjord open. The decadal cycle of the fishery which can be detected in the 1670-1750 series may also be explained as a biological pattern, while the depression of the Limfjord fishery in the late eighteenth century may have been partly caused by economic forces in the shape of the overwhelming Bohuslen exports in the 1760s, though the long-term decline which had begun already by the 1730s should be noted.

The periodicity of the Bohuslen herring implies that environmental forcing was the major underlying factor, though the sudden cessation of the fishery in 1590 was partly caused by market factors. The alternating cycles of North European pelagic stocks seems to have been correlated to the North Atlantic Oscillation, and the historical data for the Bohuslen herring stock is one of the best examples of this correlation. The most probable explanation, which parallels the experience from the nineteenth-century herring fishery on the same resource, is that the herring shoals preferred spawning grounds outside the reach of the inshore fishermen.[37] If, because of a change in salt concentration, spawning suddenly took place in the middle of the Skagerrak rather than in the sheltered archipelago of Bohuslen, the shorebound Danish and Norwegian fishermen may have found it impossible to catch the herring. In order to go deep-sea fishing they would have needed herring drifters of the Dutch *buisen* type. We do not know if and to what extent Danish fishermen had vessels of this kind, but the indication is that they did not, perhaps for lack of capital and technological expertise, including shipbuilding, skills. Thus, Dutch fishermen were able to pursue a herring fishery in the Skagerrak, while the Danish and Norwegian fishermen were not.

It is not clear when, and by how much, herring catches declined in the Sound, but it seems likely that the fishery was declining by the mid-sixteenth century. The contraction does not seem to have occurred overnight, but rather to have been an extended process. An irreversible decline after 1600 is well documented, but a careful evaluation of all estate records and other circumstantial evidence is required before the development can be fully assessed. Still, it is

---

[36]*Danske Magazin*, III, No. 4 (1999), 329-332.

[37]See Poul Holm, *Kystfolk: kontakter og sammenhœœnge over Kattegat og Skagerrak ca. 1550-1914* (Esbjerg, 1991), chapter 7.

certain that the herring in the Sound never disappeared completely, as has been asserted.

Whereas herring is a notoriously volatile resource, cod and haddock are relatively stable. Nevertheless, by the turn of the seventeenth century, the West Jutland fisheries were also in serious decline and remained adepressed after 1620. The decline was not abrupt. It seems to have begun in the second half of the sixteenth century, with tax revenues halving between 1562/1563 and about 1600, and falling by a further third by 1630. For the time being, a number of hypotheses may be put forward to explain the long-term decline but detailed research is needed on the micro-scale (which the records will allow) before the explanations can be properly tested.

The continuing depression of the Danish fisheries indicates that long-term factors at the macro-level were working in tandem with sudden and abrupt environmental changes at the micro-level. The North Atlantic Oscillation is one such macro-explanation, and others can be advanced. The cool and stormy weather prevailing in the period 1500-1850 may conceivably have altered the marine habitat and caused a long-term decline in inshore fisheries.[38] But there is no study to substantiate this possibility, and as the evidence stands, there is no immediate correlation to be found. After all, the Dutch fisheries in the North Sea were thriving while the West Jutland fisheries in the same sea were abandoned. Human and environmental factors were at play, and marine environmental historians need to join forces with historical environmental scientists to make sense of a complicated pattern.

---

[38]The period is known as the "Little Ice Age;" see *Encyclopedia Britannica* (1999 CD-ROM edition) for an updated discussion. The worst weather seems to have occurred in the latter part of the seventeenth century. An international climatological project will produce new evidence on these phenomena in the next few years (information from the Danish Meteorological Institute).

Historical Approaches to the
Northern California Current Ecosystem

R.C. Francis, J. Field, D. Holmgren and A. Strom

Abstract

This paper attempts to show how historical ecology, paleoclimatology and environmental history might come together to describe the structure and dynamics of the coastal marine ecosystem off the west coast of North America over the past several centuries. From this historical record we see that the stock of commercially and ecologically important species in the region has fluctuated in response to large-scale climate forcing. In addition, it appears that the twentieth century has been different from the late eighteenth and nineteenth centuries both climatically and biologically. This suggests that the ecosystem was structured very differently in previous centuries than it is now.

> Far from being an esoteric concern, the development of an historic sensibility ought to be considered fundamental to conservation biology.[1]

Introduction

The continental margin of North America off the coasts of California, Oregon Washington and southern British Columbia, the California Current Ecosystem (CCE), is a highly productive region that has long supported the commercial harvest of a variety of marine resources. This region is an ecotone, composed of a small number of endemic coastal species and a larger mixture of subarctic and subtropical species, many near the periphery of their distributional range.[2] It is an "open" system, meaning that it is a transition environment between subarctic and subtropical water masses and the freshwater systems that enter the ocean along its

---

[1]C. Meine, "It's About Time: Conservation Biology and History," *Conservation Biology*, XIII, No. 1 (1999), 1-3.

[2]D.L. Bottom, *et al.*, "Research and Management in the Northern California Current Ecosystem," in K. Sherman, L.M. Alexander and B.D. Gold (eds.), *Large Marine Ecosystems: Stress, Mitigation and Sustainability* (Washington, DC, 1993).

landward boundary (Figure 1). The CCE constitutes one of only five global oceanic regions where the biological production is (to a significant degree) determined by wind-driven coastal upwelling.[3] Large-scale advection of nutrient rich water from the subarctic region has been suggested as an alternative and, perhaps complementary, mechanism to explain highly significant low-frequency patterns in biological production.[4]

Figure 1
Relevant Large-scale Upper-level Physical Oceanography of the Subarctic North Pacific and Bering Sea

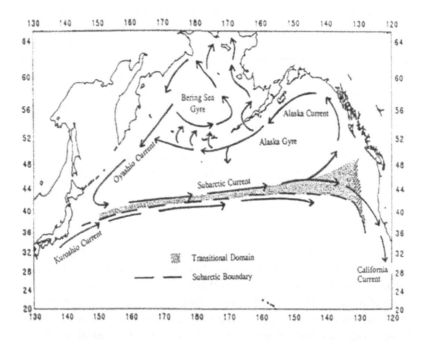

Source: B.M. Hickey, "Patterns and Processes of Circulation over the Shelf and Slope," in M.R. Landry and B.M. Hickey (eds.), *Coastal Oceanography of Washington and Oregon* (Amsterdam, 1989), 41-116.

---

[3]K.H. Mann and J.R.N. Lazier, *Dynamics of Marine Ecosystems: Biological-Physical Interactions in the Ocean* (Oxford, 1993), chapter 1.

[4]D.B. Chelton, P.A. Bernal and J.A. McGowan, "Large-scale Interannual Physical and Biological Interaction in the California Current," *Journal of Marine Research*, XL (1982),1095-1125.

The northern half of the CCE, the region of coastal ocean between Cape Mendocino (CA) and the northern tip of Vancouver Island (BC), is a zoogeographic transition between Californian and Aleutian biological provinces. Fisheries of these regions tend to be spatially extensive and fairly uniform in species content. This region (NCCE) is thus an appropriate ecological unit for regional management and is the focus of this paper. Although we focus on the NCCE, much of the paleoscientific information that we use comes from the southern part of the CCE, primarily the region off central and southern California.

Industrial-scale fishing pressure in the NCCE has been intense for only decades, although significant exploitation of one kind or another has been on-going for several centuries. At present, the most heavily fished species groupings include long-lived groundfish (e.g., rockfish, flatfish and roundfish); anadromous salmon; migratory coastal pelagics (e.g., anchovy, sardine, hake and mackerel); and several species of crustaceans (e.g., Dungeness crab and pandalid shrimp). Separating the effects of human harvest from those of environmental variability on populations is difficult, and in no case can changes be attributed solely to humankind. And yet the CCE, with its relative abundance of paleo-reconstruction studies (particularly in the Santa Barbara Basin) and short history of exploitation, affords the chance to explore the impacts of both commercial fishing and environmental variability on a valuable large marine ecosystem.

What Do We Presently Know?

Knowledge of NCCE structure and dynamics comes primarily from stock assessments, fisheries and food habits data, and oceanographic records. These sources mostly cover recent decades and indicate high levels of physical and biological variability on broad spatial and temporal scales. Catch statistics provide perhaps the most complete record of population trends and show that rapid and perhaps undesirable changes are occurring in NCCE structure. Catches of fish and marine mammals increased from the late 19th century, with major removals and consequent depletions of major predator species occurring especially in the last half of the 20th century[5].

The first species to be the subject of major commercial exploitation were marine mammals. Fisheries for sea otters, fur seals, sea lions and elephant seals grew to support enormous commercial harvests in the early and mid-1800s. While not all of these populations breed in the NCCE, all spend at least some of their time there, where many are significant predators. Estimates of actual take of most of these animals are difficult to derive. Whereas some estimates of otter and fur seal removals exist from fur trade records, the number of elephant seals and sea

---

[5] Figures illustrating the changes in catches over time are available under Publications at the HMAP website http://www.cmrh.dk/hmapindx.html.

lions taken for oil and meat can only be coarsely assessed. Charles Scammon suggested that hundreds of thousands of elephant seals must have been taken in the early part of the century, and the species was commercially extinct by the 1870s.[6]

Most coastal baleen whale and pinniped populations of the California Current were greatly reduced by the middle of the nineteenth century and continued to be depressed due to harvest or bounty hunting throughout the twentieth century. For example, specimen and museum collectors took elephant seals even as the population was on the verge of extinction.[7] Sea lions continued to be killed for the testes and penises of breeding bulls (trimmings) even after depletion. Harbour seals were targeted by fishermen and bounty hunters along the US and Canadian coastlines from the 1920s and until the 1960s; as many as several thousand animals were killed each year.[8] These removals kept most pinniped populations at low to moderate levels until the mid-twentieth century; only in the past several decades have populations of most marine mammals recovered.

Large-scale commercial harvests of fish did not begin until marine mammal populations had been significantly depleted. The salmon fisheries were among the first to develop. The first salmon cannery was opened in 1864 along the Sacramento River but was soon moved to the much more productive Columbia River. The canning business grew extremely rapidly, from 4000 cases in 1866, to over 150,000 in 1870 and 450,000 by 1876. By 1885 salmon catches in the Columbia alone topped forty million pounds, and salmon fisheries had expanded to nearly every major watershed in the Pacific Northwest.

Salmon fisheries continued to dominate the region, even as the legendary California sardine fishery began in the early twentieth century. Total sardine landings would reach over 700,000 metric tons per year, and although the vast majority were landed in southern waters, nearly 80,000 tons a year were caught in the summer fisheries of the NCCE between 1920 and 1950. When this fishery began to decline in the 1920s, it did so from north to south, suggesting that changes in marine conditions played at least some role. Landings fell first in the waters off Vancouver Island, then from Washington, Oregon and Northern

---

[6]C.M. Scammon, *The Marine Mammals of the North-Western Coast of North America; Described and Illustrated, together with an Account of the American Whale-Fishery* (San Francisco, 1874; reprint, New York, 1968).

[7]B.S. Stewart, *et al.*, "History and Present Status of the Northern Elephant Seal Population, in B.J. LeBoeuf and R.M. Laws (eds.), *Elephant Seals: Population Ecology, Behavior and Physiology* (Berkeley, 1994).

[8]P. Bonnot, *Report on the Seals and Sea Lions of California* (Sacramento, 1928); and T.C. Newby, "Changes in the Washington State Harbor Seal Population, 1942-1972" (unpublished report, B.D.).

California, until the only fishery left was in the Southern California Bight. In British Columbia, substantial herring fisheries that had developed during the period of the sardine fishery grew rapidly following the sardine's collapse.

Halibut was the first of the groundfish species to be the target of commercial fisheries, although other flatfish, such as lingcod, rockfish, and sailfish, were all supporting substantial fisheries throughout the NCCE by the turn of the century. Total landings of all groundfish and hake remained under 20,000 tons per year until the late 1960s, when the Soviet Union began fishing for hake and rockfish in the coastal waters of Oregon, Washington and British Columbia. Landings quickly reached several hundred thousand tons per year, as fishing fleets from other nations also entered the fishery. As a result of declarations of 200-mile fishing zones by both the US and Canada in the late 1870s, foreign fisheries were phased out, but overall landings continued to grow as domestic fishing fleets increased their capacities.

Since the development of a large-scale NCCE domestic groundfish fishery in the late 1870s, precipitous declines in several stocks of Pacific rockfish (*Sedates* spp.), roundfish, and flatfish have occurred and are evident in the corresponding declines of commercial landings.[9] Similarly, the salmon fisheries of the NCCE, in particular the ocean salmon fisheries, have been in decline since the late 1870s. These decreases may be attributed to a combination of unfavourable ocean conditions, loss of spawning and rearing habitat, and over-fishing.[10] Juvenile salmon, a staple in the diet of many higher trophic level species, are particularly impacted by habitat loss and interdecadal variations in climate.[11]

In contrast to the declining population trends exhibited by the NCCE salmon and groundfish fisheries over the past twenty years, the valuable crustacean fisheries of the Pacific Northwest (pandalid shrimp and Dungeness crab) exhibit extreme fluctuations over relatively short periods of time. Recent research suggests that Dungeness crab dynamics respond to both internal population feedback and large-scale environmental changes.[12] It follows that these fluctuations are due to population variability in response to a combination of

---

[9] S. Ralston, "The Status of Federally Managed Rockfish on the B.S. West Coast," in M.M. Aklavik (ed.), *Marine Harvest Refuge for West Coast Rockfish: A Workshop* (Springfield, VA, 1998), 6-16.

[10] Hare, Mantua and Francis, "Inverse Production Regimes."

[11] M.J. Mantua, *et al.*, "A Pacific Interdecadal Climate Oscillation with Impacts on Salmon Production," *Bulletin of the American Meteorological Society*, LXXVIII, No. 6 (1997), 1069-1080; and Hare, Mantua and Francis, "Inverse Production Regimes."

[12] K. Higgins, *et al.*, "Stochastic Dynamics and Deterministic Skeletons: Population Behavior of Dungeness Crab," *Science*, CCLXXVI (1997), 1431-1435.

anthropogenic influences (e.g., fishing and habitat degradation) and fluctuating oceanographic conditions.

Figure 2
Seasonal Migratory Pattern of Pacific Whiting in the California Current System, Showing the General Pattern for Several Pelagic Species

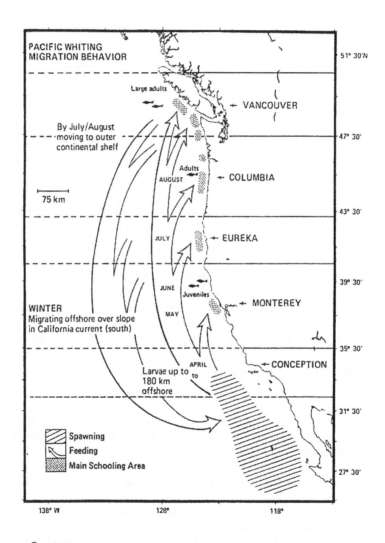

Source: See text.

Figure 3
Times Series of Summer (Frame A) and Winter (Frame B)
Abundance of Four NCCE Finfish Species Groups

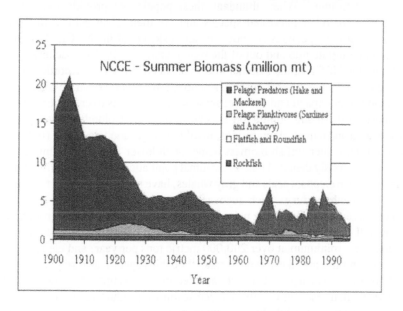

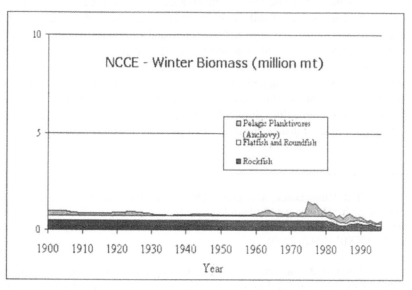

Source: See text.

Recent paleoscientific research and stock assessments show that the abundances of CCE pelagic fish species have varied significantly over periods ranging from decades to centuries, most likely in association with both climatic forcing and fishing.[13] When abundant, these populations provide a substantial forage base for other harvested species and are themselves subject to fishing pressure. But most pelagics are not constant residents of the NCCE. Both hake and sardine migrate into and out of the region annually. Pacific mackerel move north into the NCCE in the summer and south in the winter, in accordance with shifting water temperatures (figure 2). Anchovy have several separate coastal sub-populations that expand and contract seasonally, as well as over longer periods. The most northern sub-population is centred near the Columbia River plume. These spatial and temporal migrations, involving huge amounts of fish (figure 3), result in total winter finfish biomass being much lower than in the summer.

Previously depleted marine mammal populations, notably California sea lions, Northern elephant seals and gray whales, have made spectacular recoveries in the twentieth century. Such recoveries may be altering trophic dynamics through increased predation on pelagics and juvenile salmonids and groundfish. The overall biomass consumed by sea lion and seal populations alone has been estimated as a minimum at over 200,000 metric tons per year, and elephant seal consumption rates are likely even greater.[14] Seabird species also rely heavily on forage fish as a primary food source, and declines in pelagic populations in any given year often result in greatly reduced seabird reproductive success.

In summary, while little is known about the ecological structure of the NCCE, it is clear that major shifts in biomass of key trophic components occur in response to both human activities and environmental variability. These population shifts include volatile population increases and decreases (e.g., crustacea); predictable spatial migrations (e.g., pelagics); periodic species range expansions and contractions (e.g., pelagics); and steady biomass declines (e.g., groundfish, salmon and several marine mammals). In other words, dynamic species assemblages and interactions define the NCCE trophic structure, which changes over time and space. Further, the large-scale physical and biological properties of

---

[13]T.R. Baumgartner, A. Soutar and V. Ferreira-Bartrina "Reconstruction of the History of Pacific Sardine and Northern Anchovy Populations over the Past Two Millennia from Sediments of the Santa Barbara Basin," *California Cooperative Fishery Investigations Reports*, XXXIII (1992), 24-40.

[14]National Marine Fisheries Service, "Investigations of Scientific Information on the Impacts of California Sea Lions and Pacific Harbor Seals on Salmonids and on the Coastal Ecosystems of Washington, Oregon and California," NOAA Technical Memorandum NMFS-FWFSC-28 (1997).

the NCCE (e.g., temperature and flow patterns, primary and secondary production) also vary greatly over time and space with atmospheric forcing.

Paleo-Historical Studies in the NCCE: A Context for Historic Reconstruction

Paleoscientific studies in the NCCE can be roughly divided into two categories – paleoclimatology and paleoecology. Both types have shown processes with variability ranging from annual to centennial. A growing body of research shows that during the twentieth century both El Niño-Southern Oscillation (ENSO) and Pacific Decadal Oscillation (PDO) processes have had major impacts on the organization and dynamics of marine ecosystems of the northeastern Pacific and on fisheries within them.[15] Since climatic effects are so pervasive, as a first step to ecosystem reconstruction it is essential to reconstruct past patterns of climate.

The instrumental record of ENSO goes back to the mid–1800s. Historic and paleoclimatological reconstructions of ENSO now extend back to the late eighteenth century.[16] Analyses of these instrumental and reconstructed data have revealed variations in frequency modes over time. Further, the results of wavelet analysis and other analyses show the amount of "power," that is the intensity of climatic variability, at various frequencies in the ENSO time series as it progresses from 1750 to 1950.[17] The major feature that stands out is a lack of

---

[15]On the ENSO, see W.S. Wooster and D.L. Fluharty, *El Nino North* (Seattle, 1985). On the PDO, see Mantua, *et al.*, "Pacific Interdecadal Climate Oscillation." Studies of the impact of these two phenomena on fisheries include R.C. Francis, *et al.*, "Effects of Interdecadal Climate Variability on the Oceanic Ecosystems of the NE Pacific," *Fisheries Oceanography*, V, No. 1 (1998), 1-21; J.A. McGowan, D.R. Cayan and L.M. Dorman, "Climate-Ocean Variability and Ecosystem Response in the Northeast Pacific," *Science*, CCLXXXI (1998), 210-217; R.J. Beamish, "Climate and Exceptional Fish Production off the West Coast of North America," *Canadian Journal of Fisheries and Aquatic Sciences*, L (1993), 2270-2291; J.F. Piatt and P. Anderson, "Response of Common Murres to the Exxon Valdez Oil Spill and Long-term Changes in the Gulf of Alaska Marine Ecosystem," *American Fisheries Society Symposium*, XVIII (1995), 720-737; Anderson and Piatt, "Community Reorganization in the Gulf of Alaska following Ocean Climate Regime Shift," *Marine Ecology Progress Series*, CLXXXIX (1999),117-123; and A.B. Hollowed and W.S. Wooster, "Variability of Winter Ocean Conditions and Strong Year Classes of Northeast Pacific Groundfish," *ICES Marine Science Symposium*, CXCV (1992), 433-444.

[16]W.H. Quinn, D.T. Neal and S.E.A. de Mayolo, "El Nino Occurrences over the Past Four and a Half Centuries," *Journal of Geophysical Research*, XCII (1987), 14449-14461.

[17] Figures illustrating the analyses of the ENSO, PDO and related time series can be seen under Publications on the HMAP website: http://www.cmrh.dk/hmapindx.html.

power or energy in the ENSO process from about 1800 to 1880. We will return to this later.

The instrumental record of the PDO goes back only to 1900, but this has been extended in a number of PDO reconstructions using tree ring growth indices from the Northeast Pacific rim. These reconstructions suggest among other things that the interdecadal pattern which we call the PDO was not as strong prior to about 1880 and appears to be largely a twentieth-century phenomenon.

A number of paleo-oceanographic reconstructions have been made in and on the periphery of the California Current System (CCS) at the decadal-centennial scale. T.L. Jones and D.J. Kennett indicated that during the first 1300 years AD, sea surface temperatures (inferred from oxygen isotopes in mussel shells) were relatively stable (weak seasonal variation) and cooler than present.[18] In contrast, the period from 1300 to 1500 AD (late Medieval Warm period) was marked by stronger seasonal variations and warmer waters than at present. In fact, seasonal flux in sea surface temperatures may have increased to almost twice the modern range, which is inconsistent with very strong El Niño conditions, and suggestive of prolonged droughts.[19] From 1500 to 1700 AD, sea surface temperatures were once again cooler and more stable, consistent with previous isotopic studies.[20]

North of the Santa Barbara Basin in San Francisco Bay, B.L. Ingram and his colleagues used oxygen and carbon isotopic measurements of fossil bivalves contained in estuarine sediment to reconstruct an 800-year record of salinity and streamflow. They showed that between 1200 and 1950 AD (+/- 100 years uncertainty in their chronology), streamflow to San Francisco Bay was a particularly good indicator of paleoclimate in California and that this varied with a period of 200 years. Alternate wet and dry (drought) intervals lasted typically forty to 160 years, with particularly dry periods occurring between 1190-1240 AD, 1400-1460 AD and 1730-1790 AD. The first of these periods falls within the "Medieval Climatic Anomaly," a time of unusually warm and dry conditions.[21]

---

[18]T.L. Jones and D.J. Kennett, "Late Holocene Sea Temperatures along the Central California Coast," *Quaternary Research*, LI, No. 1 (1999), 74-82.

[19]S. Stine, "Extreme and Persistent Drought in California Sand Patagonia during Mediaeval Time," *Nature*, CCCLXIX (1994), 546-549.

[20]R.B. Dunbar, "Stable Isotope Record of Upwelling and Climate from Santa Barbara Basin, California," in J. Thiede and E. Suess (eds.), *Coastal Upwelling. Its Sedimentary Record. Part B: Sedimentary Records of Ancient Coastal Upwelling* (New York, 1983), 217-246.

[21]B.L. Ingram, J.C. Ingle and M.E. Conrad, "A 2000-year Record of Sacramento-San Joaquin River Inflow to San Francisco Bay Estuary, California. *Geology*, XXIV (1996), 331-334; Ingram, Ingle and Conrad, "Stable Isotope Record of Late

The paleoecological record has been developed based on fish and plankton remains extracted from oceanographic sediment cores and skeletal remains recovered from archeological sites in the Pacific Northwest. The reconstruction of abundance histories of key marine species by analysing bottom sediments at sites with annual layering and well-preserved biological remains has proven remarkably valuable.[22] This work was initiated to investigate the natural variability of the Pacific sardine population prior to its collapse under intense commercial exploitation. Several studies have used fish scales to reconstruct the natural variability of small pelagic fish populations, particularly sardines, anchovy and hake, over time scales of several decades to many centuries.[23]

Appropriate sites for the study of fish scale deposition are rare due to the suite of conditions required for preservation of the biological remains within a detailed chronological framework of deposition.[24] The critical factors are anoxic or suboxic bottom water associated with a tranquil depositional environment. Until recently, such work in the CCE was confined to two sites: Santa Barbara Basin off southern California and Soledad Basin off southern Baja California. Holmgren-Urba and Baumgartner have identified a third site within the central Gulf of California.[25] We have now successfully sampled a more northerly site, Effingham Inlet on the west coast of Vancouver Island. This site is of particular interest because it contains the remains of two coastal pelagics (sardine and hake) that

---

Holocene Salinity and River Discharge in San Francisco Bay, California," *Earth and Planetary Science Letters*, CXLI (1996), 237-247; Stine, "Extreme and Persistent Drought;" and Jones and Kennett, "Late Holocene Sea Temperatures."

[22]A. Soutar, "The Accumulation of Fish Debris in Certain California Coastal Sediments," *California Cooperative Fishery Investigations Reports*, XI (1967), 136-139; and Soutar and J.D. Isaacs, "Abundance of Pelagic Fish during the 19th and 20th Centuries as Recorded in Anaerobic Sediments off the Californias," *Fisheries Bulletin*, LXXII (1974), 257-294.

[23]Soutar and Isaacs, "Abundance of Pelagic Fish;" P.E. Smith, "Biological Effects of Ocean Variability: Time and Space Scales of Biological Response," *Rapports et Proces Verbaux, Conseil Permanent International pour l'Exploration de la Mer*, CLXXIII (1978), 117-127; Baumgartner, Soutar and Ferreira-Bartrina "Reconstruction of the History of Pacific Sardine;" and D. Holmgren-Urba and T.R. Baumgartner, "A 250-Year History of Pelagic Fish Abundances from Anaerobic Sediments of the Central Gulf of California," CalCOFI Rep. XXXIV (1993), 60-68.

[24]A. Soutar and P.A. Crill "Sedimentation and Climatic Patterns in the Santa Barbara Basin during the 19th and 20th Centuries," *Geological Society of America Bulletin*, LXXXVIII (1977), 1161-1172.

[25]Holmgren-Urba and Baumgartner, "A 250-Year History."

seasonally migrate into the NCCE at a rate which, we believe, is controlled by inter-decadal climate variability. Both are major sources of forage and targets of long-standing fisheries in the Pacific Northwest. Reconstructions from this site are providing insights on how both the abundance and distribution of these populations have fluctuated over time.

Figure 4
Coastal Pelagic Abundance of Anchovy, Sardine and Hake
Based on Paleo-reconstructions in the California Current Ecosystem

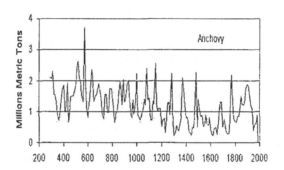

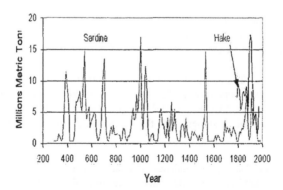

Source: T.R. Baumgartner, A. Soutar and V. Ferreira-Bartrina "Reconstruction of the History of Pacific Sardine and Northern Anchovy Populations over the Past Two Millennia from Sediments of the Santa Barbara Basin," *California Cooperative Fishery Investigations Reports*, XXXIII (1992), 24-40; and P.E. Smith, "Biological Effects of Ocean Variability: Time and Space Scales of Biological Response," *Rapports et Proces Verbaux, Conseil Permanent International pour l'Exploration de la Mer*, CLXXIII (1978), 117-127.

To illustrate the types of information recovered from paleoecological studies of fish abundance, Figure 4 shows time series plots of estimated anchovy and sardine biomass from approximately 300 AD to the present, derived from a combination of twentieth-century fishery stock assessments and paleo-reconstructions in the Santa Barbara Basin.[26] A much shorter record of hake paleo-biomass comes from the work of P.E. Smith.[27] Four points arise from analysis of these data. First, all three species tend to show interdecadal spikes of high abundance. The dominant frequencies of these spikes (sardine and anchovy) are around sixty years. Second, time series analysis indicates that sardine tend to have the most power or energy in their interdecadal variability prior to 1000 AD, while this is true for anchovy from 1000 AD to the present. Third, with the exception of a spike around 1500 AD, sardine abundance seems to have been relatively low for the entire Little Ice Age cool period from 1450 to 1850. Finally, in the late nineteenth century, coastal pelagic production of all three species (sardine, anchovy, hake) increased significantly in both intensity and variability.

The Effingham cores that we have analysed corroborate the paleoclimate reconstructions reported above, particularly the centennial scale changes in dominant frequencies and amplitudes of both the PDO and ENSO. The results indicate that if there is an association between the cadence of the PDO and the well-being of the pelagic species found in Effingham sediments, anchovy and hake did better during high frequency PDOs (nineteenth century); sardine and herring seemed to be favoured during "slow beating" PDOs (eighteenth and twentieth centuries).

Shifting to a pelagic predatory species, S.J. Crockford presents archeological evidence for the presence of adult bluefin tuna in the waters off the west coast of British Columbia and northern Washington State for the past 5000 years. Results indicate that aboriginal hunters successfully harvested large bluefin tuna. But there are no twentieth century records of adult bluefin being harvested, indicating that either the distribution of Pacific bluefin changed fairly recently, or that the specific environmental conditions favoring the movement of large bluefin into the coastal waters of the Pacific Northwest have not occurred in this century. According to ethnographic records, tuna were harpooned at night while feeding at the surface in inshore waters during unusually warm periods in late summer. Crockford concludes that it is reasonable to assume that large adult bluefin have not frequented the northern coastal waters of the eastern Pacific during the last

---

[26]Baumgartner, Soutar and Ferreira-Bartrina "Reconstruction of the History of Pacific Sardine."

[27]Smith, "Biological Effects of Ocean Variability."

century.[28] Indeed, the only recorded sightings in recent years have been on rare occasions such as during the strong El Niño event of 1957-1958.[29]

Historical information from the NCCE is included in some reports of early explorers and naturalists. For the purpose of this paper, we focus on information about the presence or absence of Pacific sardine in the NCCE. These reports suggest that three major northward expansions in the range of sardines have occurred since the late 1700s. On the first two occasions, sardines were abundant enough to support substantial fisheries that persisted for a number of years as far north as Puget Sound and Vancouver Island. Historical observations begin with the arrival of the Hezeta and Bodega y Quadra expedition in 1775 and the explorations of James Cook in 1778. The Spaniards reported sardines at Trinidad Bay in northern California, but were too weakened by scurvy to carry out thorough investigations as they travelled further north.[30] Cook's party spent most of April 1778 anchored in Nootka Sound on Vancouver Island, and Cook reported seeing large quantities of "sardines – or a small fish very much like them" caught by the local natives, but Captain Charles Clerke identified these fish as anchovies.

Shortly after Cook's expedition, a lucrative trade for sea otter pelts developed, which transformed Friendly Cove in Nootka Sound into an important commercial port.[31] This increased activity led in turn to better descriptions of local flora and fauna. Between 1786 and 1792 a number of accounts, including those of trained naturalists, described the native fisheries for sardines along the Pacific Northwest coast.[32] By the beginning of the nineteenth century, however,

---

[28]S.J. Crockford, "Archeological Evidence of Large Northern Bluefin Tuna, *Thunnus thynnus*, in Coastal Waters of British Columbia and Northern Washington," *Fisheries Bulletin*, XLV (1997), 11-24.

[29]J. Radovich, *Relationships of Some Marine Organisms of the Northeast Pacific to Water Temperatures, Particularly during 1957 through 1959* (Sacramento, 1961).

[30]F.A. Mourelle, *Voyage of the Sonora in the Second Bucareli Expedition to Explore the Northwest Coast, Survey of the Port of San Francisco, and Found Franciscan Missions and a Presidio and Pueblo at that Port* (1781; reprint, San Francisco, 1920).

[31]H.R. Wagner, *Spanish Explorations in the Strait of Juan de Fuca* (Santa Ana, CA, 1933).

[32]See, for example, A. Walker, *An Account of a Voyage to the North West Coast of America in 1785 and 1786* (Seattle, 1982); J. Strange, *James Strange's Journal and Narrative of the Commercial Expedition from Bombay to the North-west Coast of America* (Madras, 1928; reprint, Fairfield, WA, 1982); J. Meares, *Voyages Made in the Years 1788 and 1789, from China to the N.W. Coast of America* (London, 1790; reprint, New York,

explorers, naturalists and traders were no longer noting the presence of sardines in their journals, despite more extensive exploration in the Pacific Northwest region following the Lewis and Clark expedition.

Apparently, sardines did not return to the Pacific Northwest until the late 1880s. Extensive scientific investigations of fisheries resources, including those conducted by the Wilkes expedition in 1841, the Pacific railroad surveys of 1853-1857, and the investigations of the US Fish Commission in 1880-1881, failed to locate sardines in Pacific Northwestern waters.[33] The US Fish Commission reported in 1884 that sardines ranged from Chile to Cape Mendocino, California. By 1888-1889, however, this range had expanded to encompass Puget Sound, where they occurred during "the warmer part of the season, and are taken with herring and other species for market."[34] Landings of fresh sardines in 1888 were reported to be 60,000 lbs. By 1895 sardines were occurring in "large numbers" in Puget Sound and in 1902 were described as "abundant."[35] The first official records of sardines captured in Canadian waters did not occur until January 1900,

---

1967); E.J. Martinez, *Diary of the Voyage which I, Ensign of the Royal Navy, Don Estevan José Martinez, am Going to Make to the Port of San Lorenzo de Nuca, in Command of the Frigate Princesa and the Packet San Carlos, by Order of His Excellency Don Manuel Antonio Florez, Viceroy, Governor, and Captain-General of New Spain, in the Present Year of 1789* (1915; reprint, Glendale, CA, 1982); Wagner, *Spanish Explorations*; F.W. Howay, *Voyages of the Columbia to the Northwest Coast, 1787-1790 and 1790-1793.* (Portland, OR, 1990); and J.M. Mozino, *Noticias de Nutka: An Account of Nootka Sound in 1792* (1793; reprint, Seattle, 1991).

[33]C. Wilkes, Charles, *Narrative of the United States Exploring Expedition During the Years 1838, 1839, 1840, 1841, 1842* (Philadelphia, 1984); G. Suckley, *Report upon the Fishes Collected on the Survey in the Natural History of Washington Territory and Oregon: With Much Relating to Minnesota, Nebraska, Kansas, Utah, and California between the Thirty-sixth and Forty-ninth Parallels of Latitude: Being Those Parts of the Final Reports on the Survey of the Northern Pacific Railroad Route, Relating to the Natural History of the Regions Explored, with Full Catalogues and Descriptions of the Plants and Animals Collected from 1853 to 1860* (New York, 1860); and G.B. Goode, "Natural History of Useful Aquatic Animals," in Goode (ed.), *The Fisheries and Fishery Industries of the United States* (8 vols., Washington, DC, 1884), I, section 1.

[34]J.W. Collins, "Report on the Fisheries of the Pacific Coast of the United States," US Commission on Fish and Fisheries, *Report, 1888* (Washington, DC, 1892), part XVI, 3-269.

[35]D.S. Jordan and E.C. Starks, "The Fishes of Puget Sound," *Proceedings of the California Academy of Science*, 2nd series, V (1895), 785-855; and T.R. Kershaw, *Thirteenth Annual Report of the State Fish Commissioner to the Governor of the State of Washington* (Seattle, 1902).

when two specimens were collected in the Straits of Georgia near Nanaimo.[36] A lucrative fishery for sardines began in British Columbia in 1917-1918 and lasted until 1948-1949, when stocks again disappeared.[37]

Sardines returned to British Columbia in 1992, though not in sufficient quantities to warrant a resumption of large-scale commercial fisheries.[38] The return may also have been short-lived. In 1998 and 1999 a series of massive die-offs of sardines occurred along northern sections of Vancouver Island. Timing of these mass mortality events coincided with the 1998-1999 La Niña episode that caused sharp declines in coastal ocean temperatures.[39]

Temporal changes in sardine presence and intensity of fishing as well as reconstructions of the PDO time series and the intensity of energy or power in the ENSO process suggests that each of the range expansions of sardines documented since the latter part of the eighteenth century occurred during periods of increased North Pacific climatic activity.[40]The expansion of the late 1700s coincided with increased (high frequency) power in the ENSO spectrum from 1770-1800. The sardine expansion starting around 1880 coincided with increased power in both the ENSO and PDO spectra. Finally, the return of sardines to the BC coast in the 1990s coincided with a general expansion of the coastal population starting with the major North Pacific climate shift of the late 1870s and the record El Niño activity of the 1980s and 1990s.

Conclusions

How do we begin to fit all of this rather disparate information into a meaningful context for reconstructing the NCCE of the past? Let us focus on the coastal pelagics, which create a major predatory force in the NCCE during the summer. What do the paleo and historic records tell us? First, they tell us that climatic

---

[36]W.A. Clemens and G.V. Wilby, *Fishes of the Pacific Coast of Canada* (Ottawa, 1961).

[37]J.F. Schweigert, "Status of the Pacific Sardine, *Sardinops sagax*, in Canada," *Canadian Field-Naturalist*, CII, No. 2 (1988), 296-303.

[38]N.B. Hargreaves, D.M. Ware and G.A. McFarlane, "Return of the Pacific Sardine (*Sardinops sagax*) to the British Columbia Coast in 1992," *Canadian Journal of Fisheries and Aquatic Sciences*, LI (1994), 460-463.

[39]M. Drouin, "Pilchard Die-off Stirs Debate," *Pacific Fishing*, XX (March 1999).

[40] The relationships are illustrated in figures available under Publications on the HMAP website http://www.cmrh.dk/hmapindx.html.

activity (PDO, ENSO), which has a strong impact on the extent to which pelagic fish migrate into the NCCE during the summer feeding season, was more intense during the twentieth century than during the nineteenth. Climatic activity may have also been more variable during the eighteenth century. Second, by putting explorer and naturalist records together with twentieth-century fishery statistics, it appears that the first eighty years of the nineteenth century was a time of no significant sardine abundance in the NCCE. Starting around 1880, both climate and biological systems in the NCCE appear to have undergone major changes, including increased ENSO activity, strong interdecadal PDO patterns, the disappearance of bluefin tuna from the BC coast, and the surge in sardine abundance that fuelled the industrial fishery of the early twentieth century. Third, recent survey and harvest data suggest that the sardine population recovered throughout the 1990s and was once more abundant in the NCCE. Apparently this occurred in response to the climatic regime shift of the late 1870s and the increased ENSO activity that followed. Fourth, although there is currently a major summer industrial fishery for hake in the Pacific Northwest, it appears that the present abundance of hake may be relatively low in contrast to previous peaks in suggested by the paleo record. Finally, pronounced periods of boom and bust in the natural cycles of migratory coastal pelagic species, such as hake and sardines, suggest massive changes in primary and secondary production within the NCCE.

From the historical record we know that the abundance of commercially and ecologically important species in the NCCE fluctuate in response to large scale climate forcing. It appears that the twentieth century has been different from the late eighteenth and nineteenth centuries both climatically and biologically. This suggests that the ecosystem was structured very differently in previous centuries than it is now.

To investigate this conclusion further, one approach would be model the present state of the system and then reconstruct the system both 100 years ago, when marine mammals were at all time low levels and coastal pelagics were much more abundant than they are now, and 500 years ago, when temperatures were cooler and marine mammals had not yet been depleted. Such an approach should provide new insights into how interactions between fishing and physical change can alter both ecological structure and stability, and in so doing further the search for new directions in more sustainable fisheries management.

Potential for Historical-Ecological Studies of Latin American Fisheries[1]

Chris Reid

Abstract

This paper considers the importance of fishing within Latin America, and the region's contribution to the world's exploitation of marine animal populations. It reviews the pre-conquest, intra-regional and late twentieth-century export fisheries, questioning their relative absence in contemporary historiography. It then considers the historical analysis of Peruvian fisheries development in greater detail, citing the economic and political impediments to their development prior to their exponential growth in the 1950s and 1960s, and the factors underlying their collapse and volatility in the 1970s nd 1980s. Using the Peruvian case study, the paper evaluates the potential or environmental history and historical ecological studies in this region using historical time series of fishery and ecological data. It concludes that there is considerable scope to further our understanding of the historical development of living marine resource exploitation in the region, and specifically in the Humboldt Current ecosystem, through historical and ecological multi-disciplinary studies.

Introduction

The exploitation of natural resources has defined Latin American development. While the region is endowed with unique and valuable marine resources, agriculture, forestry and mining have established the strongest links with the world economy. Fisheries, though, have played an important role in transforming and defining the region during the twentieth century as that industry expanded from its initial Northern hemisphere concentration into the Southern hemisphere. However, rapid fisheries development throughout Latin America in the pursuit of economic development has often engendered and exacerbated overexploitation and

---

[1]I am grateful to Andy Thorpe and Alonso Aguilar Ibarra for their cooperation on earlier publications upon which this paper is based in part, although the usual disclaimer applies. For current purposes, Latin America is defined as including the main Latin American republics, but excluding the island economies of Cuba, Haiti and the Dominican Republic, which are normally defined as part of the region. The discussion is limited to marine capture fisheries.

conflicts between resource users without necessarily improving living standards or food security.[2]

While fishing was an important component of pre-conquest subsistence and intra-regional trade, its significance in colonial and post-independence states was relatively diminished. However, the second half of the twentieth century saw the region's ascendance as a major fisheries producer. This advance was led by the extremely rapid growth of industrial fishing in Peru, with its anchovy fishery having become the largest single-species fisheries in the world in the late-1950s. This fishery, however, had collapsed by the early-1970s due to a combination of overfishing, ineffective management and anomalous climatic conditions. Here we first review the pre-conquest, intra-regional and late twentieth century export fisheries in Latin America. We then consider the historical analysis of Peruvian fisheries development in greater detail, with the objective of increasing our understanding of past Southeast Pacific marine animal populations. Based on this development, we finally address the potential for undertaking environmental history and historical ecological studies in this region by evaluating the potential for developing historical time series of fishery and ecological data. We conclude that there is considerable scope for extending our understanding of the historical development of living marine resource exploitation in the region, and specifically in the Humboldt Current ecosystem, principally through the development of historical and ecological multi-disciplinary studies.

Fishing in Latin America: Stylized Facts

Fishing was an established part of pre-conquest subsistence and exchange in Latin America, as evidenced by the example of the Peruvian coastal plain. Archeological evidence indicates the existence of maritime economies during the early stages of migration into and through the region.[3] The region's population circa 2500-2000 BC depended heavily upon catching sea mammals and fishing using lines and beach seines, while also consuming shellfish and seabirds.[4] By the "Fluorescent Era" (*circa* 200-600 AD), sea fishing and hunting from boats using lines, nets and harpoons were important if unspecialized industries.[5] Sea fishing

---

[2]A. Thorpe, C. Reid, and A. Aguilar Ibarra, "The New Economic Model and Fisheries Development in Latin America," *World Development*, XXVIII (2000), 1689-1702.

[3]D.K. Keefer, *et al.*, "Early Maritime Economy and El Niño Events at Quebrada Tacahuay, Peru," *Science*, CCLXXXI (1998), 1883-1835.

[4]G.H.S. Bushnell, *Peru* (London, 1963), 36.

[5]J. Alden Mason, *The Ancient Civilizations of Peru* (rev. ed., London, 1971), 75.

was also important to the Inca civilization.[6] More generally, Parry suggests that fish was the main source of animal protein for pre-conquest populations.[7] It was also central to intra-regional trade. Helms argues that aquatic and marine resources contributed as much to subsistence in the pre- and immediate post-conquest period as agriculture, citing the role of trade fairs organized around the Orinoco turtle fisheries in integrating north-south trade.[8] Similarly, Hidalgo identifies the contribution of dried and smoked fish to subsistence and exchange in the south Atlantic region, although this was generally the product of river fisheries.[9] This trade engendered specialization as fish was traded for cereals and other produce. At the continent's extremity in Tierra del Fuego, limited alternatives fostered specialized nomadic fishing societies exploiting mammals, fish, shellfish and seabirds.[10]

By comparison, the exploitation of living marine resources appears to have been a relatively insignificant activity from the colonial period to the mid-twentieth century. Certainly, there were attempts to organize fishing. Parry, for example, cites the prosecution of fisheries for sea bream and conger eel off the Chilean coast by Catalan fishermen in the 1770s, whose catches were dried and exported to Peru's mining towns.[11] Chile drew on foreign expertise again at the turn of the twentieth century. Grimsby fishermen were recruited to establish a modern trawl fishery. The venture failed because "they had not, of course, a swift trawler and steam to carry fish, and there [was] no London market."[12] But such references to commercial fisheries are infrequent in the region's economic history. Post-independence states specialized in the export of temperate or tropical agricultural commodities and/or minerals, resources that were not readily

---

[6]Mason, *Ancient Civilizations of Peru*, 143-144.

[7]J.H. Parry, *The Spanish Seaborne Empire* (London, 1966), 217.

[8]M.W. Helms, "The Indians of the Caribbean and Circum-Caribbean at the End of the Fifteenth Century," in L. Bethell (ed.), *The Cambridge History of Latin America, Volume 1* (Cambridge, 1984), 46 and 55.

[9]J. Hidalgo, "The Indians of Southern South America in the Middle of the Sixteenth Century," in Bethell (ed.), *Cambridge History of Latin America*, 111.

[10]*Ibid.*, 116.

[11]Parry, *Spanish Seaborne Empire*, 217.

[12]G.F. Scott Elliott, *Chile: Its History and Development, Natural Features, Products, Commerce and Present Conditions* (London, 1911), 254-255.

available in the Northern hemisphere.[13] Even the most significant marine resource trade of the nineteenth century was closer to mining than hunting, for example, as mid-nineteenth century changes in European agriculture encouraged exploitation of guano deposits on islands off Peru's coast for use as fertilizer.[14]

In contrast, the market for fish products in the Northern hemisphere could be served from relatively near grounds.[15] Further, urbanization, the development of the railways and rising standards of living created a market for fresh fish that Southern economies could not competitively meet.[16] These circumstances prevailed up to the 1950s, when scarcity and growing demand fostered an interest in Southern fish resources. Southern hemisphere fisheries received no great stimulus from domestic markets either, with agricultural exports usually the staples of domestic diets.[17] Hostile consumer preferences and poor infrastructures inhibited the development of a mass market for fish. For example, attempts to supply frozen fish to Peru in 1936 were abandoned due to insufficient demand.[18] Inadequate refrigerated transport and storage and poor marketing was cited as factors constraining the Brazilian market in the 1950s.[19] Fish production throughout the region remains export-oriented: although Latin America is the world's largest per capita fish producer, only Africa records a lower level of per capita consumption.

---

[13]C. Furtado, *Economic Development of Latin America: Historical Background and Contemporary Problems* (2nd ed., Cambridge, 1976), 47.

[14]For discussions of the guano trade, see W.M.Matthew, *The House of Gibbs and the Peruvian Guano Monopoly* (London, 1991); P. Gootenberg, *Between Silver and Guano: Commercial Policy and the State in Post-Independence Peru* (Princeton, 1989); and Gootenberg, *Imagining Development: Economic Ideas in Peru's "Fictitious Prosperity" of Guano, 1840-1993* (Berkeley, 1993).

[15]The same argument does not apply to the exploitation of seals and whales, which are not covered in this paper.

[16]This compares with the refrigerated shipment of Argentinean beef to the UK from the 1880s.

[17]V. Bulmer-Thomas, *The Economic History of Latin America since Independence* (Cambridge, 1994), 123.

[18]B. Caravedo Molinari, "The State and the Bourgeoisie in the Peruvian Fishmeal Industry," *Latin American Perspectives*, XIV (1977), 104.

[19]R.E. Carlson, "The Bases of Brazil's Economy," in T.L. Smith and A. Marchant (eds.), *Brazil: Portrait of Half a Continent* (Westport, CT, 1972), 237.

Fishing was important in pre-conquest societies, but less so afterwards due to the greater opportunities offered by other economic activities. Perhaps more properly, this was a failure to develop modern specialized commercial fisheries in the European and North American fashion rather than the absence of fishing per se. Most fisheries in the region have been small scale artisanal fisheries, often conducted by socially excluded non-Spanish speaking indigenous populations, as observed by Murphy in Peru after World War One.[20] Official indifference or inadequate resources appear to have contributed to the under-recording of these activities. For example, although Peru's Servicio de Pesquería a subdivision of the Ministry of Agriculture – was nominally in charge of both marine and freshwater fisheries, it almost entirely devoted itself to the latter.

Figure 1

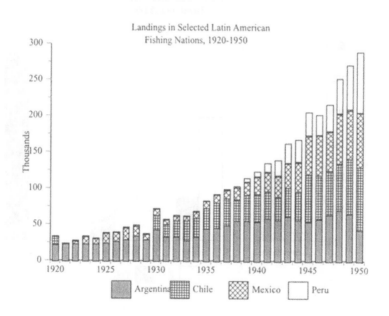

Source: Derived from United Nations Food and Agriculture Organization (FAO), *Yearbook of Fisheries Statistics* (Rome, various years).

---

[20]R.C. Murphy, "The Guano and the Anchovy Fishery," in M.H. Glantz and J.D. Thompson (eds.), *Resource Management and Environmental Uncertainty: Lessons from Coastal Upwelling Fisheries* (New York, 1981), 97.

Commercial fisheries began on a significant scale in the first half of the twentieth century. Figure 1 shows marine fish catches by Argentina, Chile, Mexico and Peru between World War One and 1950. Despite missing observations in some years and areas, it is evident that production in each country increased substantially over the period. Even so, production was modest by Northern hemisphere standards, contributing a mere 1.2% of world fish production in 1938.[21] Trade in fish products continued to be unimportant in this period – many countries apparently ran a persistent deficit in the fish trade – suggesting that fish was relatively unimportant in regional diets.

Figure 2

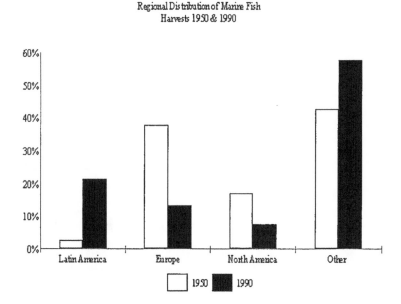

Regional Distribution of Marine Fish
Harvests 1950 & 1990

Source: See figure 1.

---

[21]J.P. Cole, *Latin America: An Economic and Social Geography* (London, 1965), 159.

As figure 2 shows, Latin America accounted for only 2.65% of world fish production by 1950.[22] Although the region's fisheries were clearly underdeveloped at this stage, their expansion was not automatic. Small industries such as fishing had limited call upon scarce development resources.[23] As Frank's postwar survey concluded:

> [T]here is a general shortage of all types of gear, the number of boats is inadequate, existing fishing fleets are in need of repairs, operations are curtailed by the shortage of gasoline and fuel oils, refrigeration and storage facilities are very limited, transportation facilities to inland areas are poor or nonexistent...fishing operations are generally conducted on a small scale, and there are not enough large-scale producing units.[24]

Further, inadequate scientific and economic information were continuing constraints.[25] Governmentally, in most instances fishing remained the unwelcome responsibility of agriculture ministries, and comparatively few resources were allocated to research or enumeration.

Despite these constraints, many Latin American states achieved an outstanding level of fisheries development after World War Two. As figure 2 shows, marine fish landings within the region during 1990 represented about one-fifth of the world's harvest, significantly exceeding catches in Europe and North America. Production rose by an average of some 9.6% per annum over four decades, an exceptional rate of growth. Christy identifies a common model for the region's fisheries development. He argues that Latin fisheries have typically passed through the successive phases of neglect, development through large-scale public sector operations, private sector development, and fisheries management.

---

[22]United Nations Food and Agriculture Organization (FAO), *Yearbook of Fisheries Statistics* (Rome, various years).

[23]Carlson, "Bases of Brazil's Economy," 237, makes this point with respect to Brazil.

[24]M. Frank, "The Fishing Industry," in L.J. Hallett (ed.), *Industrialization of Latin America* (New York, 1946), 139-141.

[25]B.F. Osorio-Tafall, "Better Utilization of Fisheries Resources in Latin America," *FAO Fisheries Bulletin*, IV, No. 3 (1951), 22.

As in most developing fisheries, the expansion of commercial exploitation greatly exceeded advances in science or management.[26]

Table 1
Distribution of Harvests from marine Fisheries in Latin America
1938, 1950 and 1990

	1938	1950			1990			
	Tonnes	Tonnes	Percent of Latin Catch	Percent of World Catch	Tonnes	Percent of Latin Catch	Percent of World Catch	Average Growth 1950-1990[1]
Argentina	55,300	39,000	10.49	0.28	502,624	3.41	0.73	6.60
Belize	300[2]	100	0.03	†	418	0.00	†	3.64
Bolivia	700	-	-	-	3976	0.03	†	-
Brazil	103,300[3]	104,500	28.11	0.74	509,165	3.45	0.74	4.04
Chile	30,600	70,300	18.91	0.50	5,014,584	34.02	7.27	11.26
Colombia	10,000	1300	0.35	†	77,207	0.52	0.11	10.75
Costa Rica	1450	400	0.11	†	11,460	0.08	0.02	8.75
Ecuador	1800	5600	1.51	0.04	303,581	2.06	0.44	10.50
El Salvador	100	700	0.19	†	2321	0.02	†	3.04
Guyane	100	200	0.05	†	2785	0.02	†	6.81
Guatemala	100	-	-	†	1666	0.01	†	
Guyana	2000	500	0.13	†	32,208	0.22	0.05	10.97
Honduras	100	100	0.03	†	4510	0.03	†	9.99
Mexico	18,700	11,200	3.01	0.08	1,012,176	6.87	1.47	11.92
Nicaragua	100	400	0.11	†	1135	0.01	†	2.64
Panama	700	400	0.11	†	129,376	0.88	0.19	15.54
Peru	4800[3]	73,500	19.77	0.52	6,776,783	45.98	9.82	11.97
Surinam	400	800	0.22	†	5240	0.04	†	4.81
Uruguay	3600	3200	0.86	0.02	89,551	0.61	0.13	8.69
Venezuela	21,700	59,500	16.01	0.42	258,657	1.75	0.37	3.74
Total	255,850[4]	371,700	-	-	14,739,423	-	-	-

Note: † - Less than 0.01 percent; [1]Average annual compound rate of growth, 1950-1990; [2]Observed 1942; [3]Observed 1939; [4]Calculated as if all observations taken in 1938.

Source: See figure 1.

---

[26]F.T. Christy, "The Development and Management of Marine Fisheries in Latin America and the Caribbean," Inter-American Development Bank Policy Research Paper, ENV-110 (1996), 19.

What caused this growth? Changes in demand were certainly significant. Fishing in the northern hemisphere reached or exceeded the limits to growth after World War Two as demand for fish products reached new heights. Of particular importance was the growing international fishmeal market, which created a demand for plenteous but previously unmarketable small pelagic species. On the supply-side, new technology and institutions encouraged developing countries to look at marine resources as a basis for economic development. Latin American states played a prominent role in establishing the new law of the sea.[27] Mexico's reaction to the 1945 Truman Proclamation in claiming national rights to marine resources bounded within the adjacent continental shelf precipitated further claims in the region.[28] This created new opportunities for fisheries development, as epitomized by the establishment of Mexico's tuna fisheries in the 1970s, but also added responsibilities that poor states often struggle to meet.[29]

How have landings been distributed within the region during the process of development? Table 1, which decomposes the region's marine fish production for 1938, 1950 and 1990, suggests that there have been significant changes over this period. In the late-1930s, Brazil, Argentina and Chile cumulatively accounted for three-quarters of the region's fish production. Only Mexico among the Central American states stands out as a fishing nation of any stature before World War Two. Care should be exercised in interpreting estimates of pre- and post-World War Two production, however. According to Osorio-Tafall's survey, fisheries statistics in Latin America at this time were "incomplete, fragmentary, and in some cases completely nonexistent."[30] Figures for most states should be regarded as nominal estimates of catches by artisanal fishers. Catches by foreign vessels –

---

[27]See Vicuña F. Orrego (ed.), *The Exclusive Economic Zone: A Latin American Perspective* (Boulder, 1984); Vicuña F. Orrego, "Trends and Issues in the Law of the Sea as Applied in Latin America," *Ocean Development and International Law*, XXVI (1995), 93-103; and F. Paolillo, "The Exclusive Economic Zone in Latin American Practice and Legislation," *Ocean Development and International Law*, XXVI (1995), 105-125.

[28]The most important of these were by Chile and Peru in 1947, which declared 200-mile "territorial" seas that afforded exclusive access to major fisheries. These claims were formalized by Chile, Ecuador and Peru in August 1952 under the Santiago Declaration, which advanced ecological, economic and political claims to resources. See C.R. Bath, "Latin American Claims on Living Resources of the Sea," *Inter-American Economic Affairs*, XXVII, No. 4 (1974), 71.

[29]See F. Castro y Castro, "Importance of the Exclusive Economic Zone to the Tuna and Fisheries Development of Mexico," in E.L. Miles (ed.), *Management of World Fisheries: Implications of Extended Coastal State Jurisdiction* (Seattle, 1989), 227-235.

[30]Osorio-Tafall, "Better Utilization," 6.

mainly from the United States – often exceed domestic production. Landings were more diversified in 1950, when the three largest producers – Brazil, Chile and Peru – accounted for about two-thirds of the region's harvest, with Venezuela and Argentina contributing a further quarter of production. Many states appeared to have a lower level of production than before the war, including Argentina and Mexico. The outstanding achievement in this period was the development of the Peruvian fishing industry, which increased production by an average of some twenty-five percent per annum between 1938 and 1950.

Landings had become vastly greater and more concentrated by 1990, when Chile and Peru alone took four-fifths of the region's catches, equivalent to approximately seventeen percent of world production. Production in many other states – Argentina, Brazil, Ecuador, Mexico and Venezuela – was greater than that achieved in many established Northern hemisphere fishing nations. But still fishing remains a marginal activity in the region. Approximately 2200 communities depended on fishing in the early-1990s, with direct employment estimated at about one million persons, about 0.25% of the region's population. The region's fisheries are heterogeneous with respect to the composition of catches, scale, technological sophistication of exploitation, and the utilization and consumption of landings, but artisanal fishers account for approximately ninety percent of direct employment.[31] These fishers mostly participate in inshore fisheries; the most significant are for shrimp, which are among the most important in terms of their contribution to local incomes and subsistence and among the most threatened by coastal development.[32]

In the early-1950s, when Osorio-Tafall argued for the better utilization of the region's fisheries resources,[33] most species were under exploited, and fishing's contribution to employment and living standards was modest. Fifty years later, the region presents a more diverse picture. Production is export-oriented, and the great majority (by volume) is for animal feed rather than for human consumption. Employment has increased, but fishing remains a marginal occupation generating low incomes. Privatization and deregulation of state-owned fisheries have encouraged industrial concentration and created powerful interests in the sector, ensuring that the economic benefits derived from living marine

[31]A. Bermudez and M. Ageuro, "Socioeconomic Research on Fisheries and Aquaculture in Latin America," in A.T. Charles, *et al.* (eds.), *Fisheries Socioeconomics in the Developing World* (Ottawa, 1994), 38-39.

[32]M.H. Lemay, "Coastal and Marine Resources Management in Latin America and the Caribbean," Inter-American Development Bank Technical Study, ENV-129 (1998), 9.

[33]Osorio-Tafall, "Better Utilization."

resources are not evenly disbursed.[34] The region's governments are now responsible for vast exclusive economic zones for which adequate management resources do not exist.[35] Many key fisheries are exploited at or beyond sustainable levels. It is unlikely that these problems will be resolved in the near future given the current levels of governmental interests and available resources.

From General to Specific: Peruvian Fisheries

The region's most important commercial fisheries are found in the upwelling Eastern Pacific boundary (Humboldt) current. This cool nutrient rich current supports large stocks of small pelagic species, the most commercially important of which are the anchovy (*Engraulis ringens*) and pilchard (*Sardinops sagax*).[36] These fisheries have been widely discussed on account of their rapid development, magnitude, commercial importance, and sensitivity to environmental change. Interest in this last characteristic currently dominates on account of the close association between the fisheries' and the El Niño climatic anomaly. Peru established the region's first industrial fishery based upon producing anchovy fishmeal for export, providing a model for Chile's subsequent fisheries development.[37]

Figure 3 shows Peru's total catch between 1950 and 1995. Peruvian firms caught about 70,000 tonnes of marine fish in 1950. There was an extremely rapid increase in landings between the mid-1950s and late-1960s, equivalent to an average rate of growth of some twenty-seven percent per annum, culminating in a peak catch of some 12.5 million tonnes in 1970. Landings decreased substan-

---

[34]The distributional consequences of fisheries development are discussed in J. Peña-Torres, "The Political Economy of Fishing Regulation: The Case of Chile," *Marine Resource Economics*, XII (1997), 253-280; A.A. Ibarra, C. Reid and A. Thorpe, "The Political Economy of Marine Fisheries Development in Peru, Chile and Mexico," *Journal of Latin American Studies*, XXXII (2000), 503-527; and Thorpe, Reid and Ibarra, "New Economic Model."

[35]See D. de G. Griffith, *et al.*, "Fisheries and Aquaculture Research Capabilities and Needs in Latin America: Studies of Uruguay, Argentina, Chile, Ecuador, and Peru," World Bank Technical Paper (Fisheries Series) 148 (1991).

[36]*Sardinops sagax* are described as either pilchard or sardine in both primary and secondary sources, although the former designation is preferred here.

[37]There were substantial differences between the Chilean and Peruvian modes of development, with the former more directed and geographically concentrated. See C.N. Caviedes, "The Impact of El Niño on the Development of the Chilean Fisheries," in Glantz and Thompson (eds.), *Resource Management*, 356-358.

tially in the early-1970s, averaging just over one-quarter of their highest value between the mid-1970s and mid-1980s, and only exceeded five million tonnes again in 1986. Over the course of the following decade, landings grew at close to five percent per annum, approaching their historic peak by the early-1990s. The pattern of growth is revealed more clearly in figure 4, which examines the growth rate of total Peruvian landings using the first differences of the natural logarithm transformation. This illustrates the exponential trend in the growth of the fisheries up to the early-1960s. There is a tendency for the trend to become more variable and to change direction post-1960, following the recurrent El Niño events. The most important of these for local fisheries development occurring in 1964-1965, 1972-1973, 1976-1977, 1982-1983, 1986-1987, 1991-1992 and 1993-1994. Although figure 4 does not present a precise association between changes in the growth of landings and El Niño events, it does suggest a correspondence between climatic change and a decrease or reversal in the rate of growth of landings in the following period(s). This is most evident following the 1972-1973 and 1982-1983 events.

Figure 3

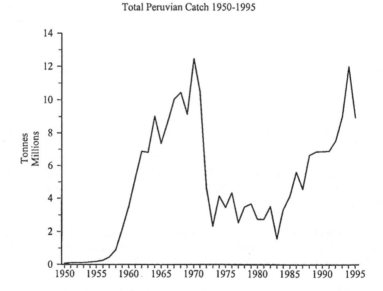

Total Peruvian Catch 1950-1995

Source: FAO, *Yearbook of Fisheries Statistics* (Rome, annual).

Figure 4

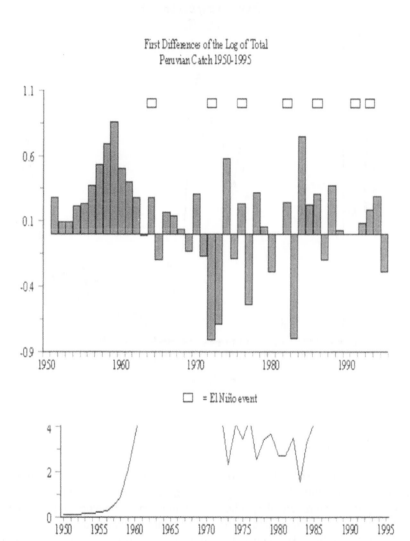

First Differences of the Log of Total
Peruvian Catch 1950-1995

☐ = El Niño event

Source: See figure 1.

Figure 5

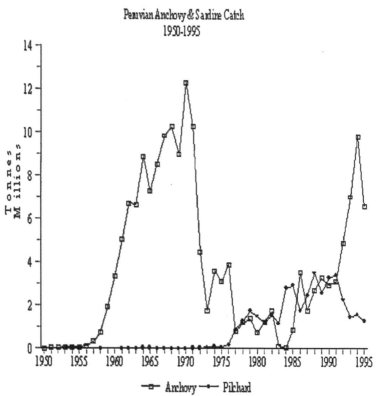

Peruvian Anchovy & Sardine Catch
1950-1995

Source: See figure 1.

A characteristic feature of Peru's fishing industries has been their
dependence on a limited range of species. Figure 5 shows that anchovy was the
most important commercial species in the industry's development up to the
early-1970s. At the peak of landings in 1970, anchovy accounted for some 98.5%
of all fish production. This concentration on a single species was unprecedented
among the world's major fishing industries. It was also shown to be a major
structural weakness when the combined effects of overfishing and the 1972-1973
El Niño event brought about a collapse of the anchovy stocks, followed by years
of industrial decline and economic hardship. Figure 5 also shows that pilchard
catches were modest and relatively unimportant until the collapse of the anchovy
fisheries. The 1972-1973 El Niño event resulted in a latitudinal redistribution of
anchovy stocks, and their replacement by pilchard and, to a lesser extent, jack

mackerel.[38] On average, pilchard catches during the 1950s and 1960s were less than 10,000 tonnes per annum. They increased rapidly after the 1972-1973 El Niño event, exceeded three million tonnes per annum by 1979, and averaged approximately 4.25 million tonnes per annum during the 1980s. This species switching ensured that the Peruvian fishing and fishmeal industries were not rendered wholly redundant by the collapse of the anchovy fishery.

The initial development of the fisheries can be attributed to Peru's links with the US economy. US firms, excluded from their normal sphere of operations, established Peru's first canneries in 1939 to produce bonito for export.[39] Peruvian capital was mobilized and there was a significant increase in the number of fishing firms, and concomitant investments in the shipbuilding industry.[40] Fishmeal production and anchovy catches were modest in the early-1950s: fishmeal was produced from the bonito canneries' waste, while anchovy catches that might be used for fishmeal directly were less than fifteen percent of total landings in 1952.[41] However, exogenous factors in the early-1950s strongly promoted the development of an anchovy fishmeal industry. On the demand-side, Japanese competition routed Peruvian canned fish products from US markets. On the supply-side, the collapse of the Californian sardine fishery provided a source of cheap capital equipment and expertise to develop a fishmeal industry.[42]

Anchovy fishing was strongly opposed by the Guano Administration Company, which feared competition between the seabirds and fishermen for a common resource, the anchovy. Murphy's 1954 report clearly articulated its views, arguing that fishers would overexploit anchovy stocks to the detriment of guano-producing seabird populations.[43] Conversely, Murphy argued that the Company's monopoly had protected and nurtured these resources. The establishment guano lobby was able to secure a restriction on anchovy fishing, ensuring

---

[38]Jack mackerel catches are not shown in figure 5 to avoid confusion.

[39]The bonito was exported to the US labeled as tuna. Osorio-Tafall, "Better Utilization," 15; and M. Roemer, *Fishing for Growth: Export-led Development in Peru, 1950-1967* (Cambridge MA, 1970), 81.

[40]Caravedo Molinari, "Peruvian Fishmeal Industry," 105.

[41]D.P. Werlich, *Peru: A Short History* (Carbondale, IL, 1978), 260-261; and *FAO, Yearbook of Fisheries Statistics*, various years.

[42]E. Ueber and A. MacCall, "The Rise and Fall of the California Sardine Empire," in M.H. Glantz (ed.), *Climate Variability, Climate Change, and Fisheries* (Cambridge, 1992), 43-44.

[43]Murphy, "Guano and the Anchovy Fishery."

that the first fishmeal plant imported from California in 1953 was erected illegally.[44] Anchovy catches increased considerably when restrictions were lifted in 1959 and the balance of power between the guano and fishmeal industries was permanently altered. Production was stimulated by the rapid growth of the international fishmeal market.[45] Rising catches were achieved using modern open sea purse seining, which Peru adopted before many established Northern hemisphere fishing nations.[46] Recognition of the fishery's importance and of the possibility of overfishing saw the establishment of the Instituto del Mar del Perú (IMARPE) in 1960 to investigate the country's living marine resources.[47]

The industry did not settle into a long-run steady state. Rather, growth became more erratic following the 1964-1965 El Niño event.[48] While this event was relatively mild, the government followed IMARPE's recommendation and introduced a month-long closed season (*veda*) in August 1965. Far from stabilizing the industry, this encouraged more intensive harvesting when the fishery was reopened, exacerbating the tendency towards overproduction.[49] IMARPE's panel of experts indicated that recorded catches during the 1969-1970 season of 9.5 million tonnes exceeded the anchovy fishery's maximum sustainable yield by 1.5

---

[44]M. Glantz, *Currents of Change: El Niño's Impact on Climate and Society* (Cambridge, 1996), 29.

[45]E. Leoncio Segura, "An Econometric Study of the Fish Meal Industry," FAO Fisheries Technical Paper, 119 (1973).

[46]Roemer, *Fishing for Growth*, 83. Power block purse seining was introduced to the herring fisheries of Iceland in 1957, Norway in 1961, and the UK in 1966. Its adoption in these established fisheries appears to have been more problematic than in Peru's infant industry.

[47]Roemer, *Fishing for Growth*, 68.

[48]Although the 1964-1965 anomaly produced relatively little fluctuation in anchovy stocks, it had a pronounced impact on seabird populations. Seabird populations had been as high as eighteen million birds in 1962, but fell to four million after the mid-1960s El Niño and did not recover. D.C. Duffy, "Environmental Uncertainty and Commercial Fishing: Effects on Peruvian Guano Birds," *Biological Conservation*, XXVI (1983), 231; and D.H. Cushing, *Climate and Fisheries* (London, 1982), 288.

[49]B. Smetherman and R.M. Smetherman, "Peruvian Fisheries: Conservation and Development," *Economic Development and Cultural Change*, XXI (1973), 345.

million tonnes.[50] In actuality, losses during fishing and vessel unloading, dumping at sea, and under-reporting, ensured that catches were nearer thirteen to fourteen million tonnes.[51] Eliminating overcapacity, the Panel argued, would increase current and future incomes and conserve the anchovy stocks. Unfortunately, as Caviedes and Fik conclude, "words of warning were heeded neither by the greedy industry owners nor by the government which was reaping the benefits from rising exports and accruing tax revenues."[52] The government was also concerned to maintain employment in the sector, which amounted to some 20,000 fishers and 3000 factory workers.[53] Opportunities for managing the fishery diminished following the 1968 military coup which fragmented policy making and weakened ties with external scientific experts to meet nationalist imperatives.[54]

Anchovy resources were therefore substantially overfished before the 1972-1973 El Niño, the immediate consequences of which were disguised by the movement of anchovy stocks inshore where they were more easily fished. Ultimately, the combination of overfishing and climatic anomaly caused the adult anchovy stock to collapse. The government responded by closing the fishery between July 1972 and March 1973, initially compensating fishers and processors.[55] However, the 1971 Fisheries Law permitted the government to

---

[50]Instituto del Mar del Perú (IMARPE), "Panel of Experts' Report (1970) on the Economic Effects of Alternative Regulatory Measures in the Peruvian Anchovy Fishery," in Glantz and Thompson (eds.), *Resource Management*, 374.

[51]G.J. Paulik, "Anchovies, Birds, and Fishermen in the Peru Current," in Glantz and Thompson (eds.), *Resource Management*, 37.

[52]C.N. Caviedes and T.J. Fik, "Modeling Change in the Peruvian-Chilean Eastern Pacific Fishery," *Geojournal*, XXX (1993), 370.

[53]M.A.F. Ros-Tonen and J.H. van Boxel, "El Niño in Latin America: The Case of Peruvian Fishermen and North-East Brazilian Peasants," *European Review of Latin American and Caribbean Studies*, LXVII (1999), 12.

[54]L.A. Hammergren, "Peruvian Political and Administrative Responses to El Niño: Organizational, Ideological, and Political Constraints on Policy Change," in Glantz and Thompson (eds.), *Resource Management*, 330-332.

[55]L.K. Boerema and J.A. Gulland, "Stock Assessment of the Peruvian Anchovy (*Engraulis ringens*) and Management of the Fishery," *Journal of the Fisheries Research Board of Canada*, XXX (1973), 2234.

Chris Reid

nationalize the anchovy fleet and processing plants,[56] even though their debts were double their capital value.[57] While nationalization and the creation of the state enterprise PescaPeru were not the only options open to the government, they were the most ideologically acceptable.[58] Nationalization and a brief flirtation with participatory ownership quickly "turned what had been a major industry providing tax revenue to a sector requiring years of massive state subsidies."[59] Although the state's divestiture of surplus vessels and plants began in 1975 its continued involvement in the industry impeded structural adjustment, worsening the sector's vulnerability to fluctuations in anchovy stocks. Even though 1976's catch promised the fishery's revival, continued fears of further stock collapse following the 1976-1977 El Niño forced the fishery's closure several times during 1977.[60]

With anchovy fishing closely regulated, private boats began to fish for pilchards and jack mackerel, catches of which increased from 54,000 tonnes to 387,000 tonnes between 1976 and 1979.[61] While anchovy had been the dominant small pelagic species off the Peruvian coast prior to the 1972-1973 event, its collapse allowed a large increase in pilchard abundance.[62] Redirecting effort did not resolve overfishing but extended it to new fisheries. In late-1980, the Ministry announced the pilchard fishery's closure in order to "manage pilchard stocks more

---

[56]The 1971 Fisheries Law established state control of marketing, provisions for displacing foreign ownership, and the introduction of worker participation. R. Thorpe and G. Bertram, *Peru 1890-1977: Growth and Policy in an Open Economy* (Basingstoke, 1978), 302.

[57]W. Royce, *Fishery Development* (Orlando, 1987), 106.

[58]Hammergren, "Peruvian Political and Administrative Responses," 339.

[59]D. Weidner and D. Hall, *World Fishing Fleets: An Analysis of Distant-water Fleet Operations Past-Present-Future. Vol. IV: Latin America* (Silver Spring, MD, 1993), 430.

[60]Anon., "Peru Closes Anchovy Fishery," *Marine Fisheries Review*, XXXIX, No. 8 (1977), 32.

[61]*FAO, Yearbook of Fisheries Statistics*, various years.

[62]K.R. Patterson, J. Zuzunaga and G. Cárdenas, "Size of the South American Sardine (Sardinops sagax) Population in the Northern Part of the Peru Upwelling Ecosystem after Collapse of the Anchoveta (*Engraulis ringens*) Stocks," *Canadian Journal of Fisheries and Aquatic Science*, XLIX (1992), 1762-1763.

carefully than the decimated anchovy stocks were managed."[63] It re-opened in 1981 with an intricate system of controls designed – unsuccessfully – to limit catches to one million tonnes. Jack mackerel stocks were not regulated, as the anchovy fleet was poorly suited to fishing this species in deeper offshore waters.

The 1980s saw a new appreciation of the need to further reduce capacity, especially after the severe 1982-1983 El Niño. However, this was economically and politically problematic, frustrating attempts to redirect production from fishmeal to human consumption. The effect of reduced fishing opportunities and inadequate incomes was clearly evident in the industrial fleet in the mid-1980s. Over three-quarters of the fleet was more than twenty years old and in poor repair.[64] Although restrictions on anchovy and pilchard fishing between early-1986 and late-1988 inhibited investment, production began to increase.

The recovery of the fisheries was arguably facilitated by the 1990 structural adjustment program, which encouraged exports and new capital formation.[65] Investment in vessel and plant improvements between 1991 and 1995 totaled some US $400 million as production returned to early-1970s levels.[66] Anchovies and pilchards continued to account for nearly ninety percent of Peruvian catches in the mid-1990s, even though their fisheries were respectively defined as "recovering, fully- to over-exploited," and "depleted, fully- to over-exploited."[67] Although Peru's fisheries remained vulnerable to climatic anomalies, the 1990s brought signs of new approaches to fisheries management to overcome the economic and political causes of boom and bust. First, there was the passing of a new General Fisheries Law in 1994 that moved Peru's fisheries

---

[63]Anon., "Peruvian Fisheries Developments, 1980-81," *Marine Fisheries Review*, XLIII, No. 7 (1981), 27.

[64]A. García Mesinas, "Current Status and Strategies of Peruvian Fisheries," in M. Murphy (ed.), *Maximum Sustainable Yield from Fish Stocks: A Challenge to Fishermen and Managers* (Cork, 1993), 82.

[65]A.A. Ibarra, C. Reid and A. Thorpe, "Neo-liberalism and its Impact on Overfishing and Overcapitalization in the Marine Fisheries of Chile, Mexico and Peru," *Food Policy*, XXV (2000), 599-622; and Thorpe, Reid and Ibarra, "New Economic Model."

[66]Anon., *WorldFish Report* (1995), SP/4; and Anon., *WorldFish Report* (1997), SP/1.

[67]*FAO, Review of the State of World Fishery Resources: Marine Fisheries* (Rome, 1997), table XV.

into Christy's "fisheries management" stage.[68] Second, the government began to conduct economic policy by rules and targets, not discretion. If this logic were extended to fisheries management, it should signal greater commitment by government for recommended total allowable catches, although powerful industry elites appear to exert an increasing countervailing influence over fisheries policy as in Argentina and Chile.[69] Third, there appeared to be a renewed commitment to fishing for human consumption, as evidenced by the opening of the world's largest fully automated fish cannery in 1997 at a cost of US \$35 million,[70] even though production remained primarily export oriented. In the future, it is possible that increased awareness of the significance of El Niño may encourage the development of ecotourism in the region, thereby increasing the economic value of guano bird populations and creating incentives to reduce fishing pressure.[71] However, new indebtedness within the fisheries sector engendered by neo-liberal reforms during the 1990s, together with repercussions from the collapse of Fujimori's neo-liberal regime, ensure that the sector remains characterized by uncertainty at the beginning of the twenty-first century.

Potential for Historical and Ecological Studies of the Humboldt Current System

The main objectives of History of Marine Animal Populations (HMAP) initiative are threefold: to expand our understanding of the long-run exploitation of living marine resources, to consider the factors underpinning changes in resource abundance, and to evaluate the role of the exploitative industries in historical development. It is an attempt to foster exchange between the sciences and the humanities, emphasizing innovations in the use of evidence and methodology.

Although further historical research into Latin America's fishing industries and cultures sector is highly desirable, HMAP expressly focuses on furthering our understanding of the state of marine ecosystems before and after intensive human exploitation began. Since the program is geared towards

---

[68]The details of the law are outlined in Ibarra, Reid and Thorpe, "Neo-liberalism and its Impact on Overfishing and Overcapitalization."

[69]Peña-Torres, "Political Economy of Fishing Regulation;" and Thorpe, Reid and Ibarra, "New Economic Model."

[70]D. Gillespie, "New Peruvian Canning Factory is 'World's Largest,'" *Seafood International* (November 1997), 35.

[71]Duffy, "Environmental Uncertainty and Commercial Fishing," 235, partially anticipated this possibility, suggesting that the portion of the total allowable catch allocated to the guano bird population could act as a buffer for the fishing industry in El Niño years.

understanding large scale change on an ecosystem-wide basis, three factors suggest that the most suitable candidate for further enquiry among the region's fisheries are those in the Humboldt current ecosystem, incorporating Peru's commercial fisheries and also those of Chile. First, the Southeast Pacific has hosted the largest commercial fisheries of any historical epoch as measured by landings, exerting a powerful influence during the course of world fisheries development. Second, the development of these fisheries was explosive and intensive, conditioned by the adoption of radical fishing technologies within new territorial limits. This contrasts with the more gradual evolution of the Northern Hemisphere fishing industries during the nineteenth and early-twentieth centuries through the extension of fishing grounds. Finally, the region's fisheries are important because of the influence of climatic anomalies. Besides raising important questions regarding the relative significance of environmental forcing and economic, social and political forces in fisheries development, this also provides a common link with pelagic fisheries in other upwelling systems, such as the California and Benguela currents.[72]

Previous studies of the Humboldt Current fisheries have addressed a wide range of hypotheses. Caviedes and Fik reduce these to three fundamental hypotheses that attribute the variance of the fisheries to changes in oceanic-climatic conditions, to human predation, and to long-term regime shifts, respectively.[73] Various studies have tested these hypotheses using univariate and multivariate time series analysis with some success.[74] Differences in hypotheses, methodologies and techniques, data utilized, and the spatial and temporal limits of various studies, prohibit any overarching summary, but there are common issues that are of interest and concern for historians.

---

[72]Several recent studies have undertaken comparative long-run analyses of these systems to explore the existence of "regimes" – long-term regular oscillations in the abundance. See, for example, D. Lluch-Belda, *et al.*, "World-wide Fluctuations of Sardine and Anchovy Stocks: The Regime Problem," *South African Journal of Marine Science*, VIII (1989), 195-205; and L.B. Klyashtorin, "Long-term Climate Change and Main Commercial Fish Production in the Atlantic and Pacific," *Fisheries Research*, XXXVII (1998), 115-125.

[73]C.N. Caviedes and T.J. Fik, "The Peru-Chile Eastern Pacific Fisheries and Climate" in M.H. Glantz (ed.), *Climate Variability, Climate Change, and Fisheries* (Cambridge, 1992), 356-357.

[74]See, for example E. Yáñez, M.A. Barbieri and L. Santillán, "Long-term Environmental Variability and Pelagic Fisheries in Talcahuano, Chile," *South African Journal of Marine Science*, XII (1992), 175-188; Caviedes and Fik "Peru-Chile Eastern Pacific Fisheries;" Caviedes and Fik "Modeling Change in the Peruvian-Chilean Eastern Pacific Fishery;" and Klyashtorin, "Long-term Climate Change."

While historians are unusually preoccupied by sources, the availability and robustness of data is a universal concern. With respect to Peruvian sources, the enumeration of the fisheries lacked a guiding institution during their development. Various agencies – including the Ministry of the Navy, the Ministry of Agriculture's Fisheries Service and the Council of Hydrobiological Research – bore some responsibility for fisheries research before the establishment of IMARPE. It would appear that data for the 1940s and 1950s reported by the FAO was collected primarily to inform the Ministry of Finance which exerted a strong influence over the fishing industry.[75] The establishment of IMARPE in 1960 ensured that consistent estimates for catch and effort were available from an early stage of the industry's development.[76] Therefore, while it is possible to present a long-run time series of aggregate Peruvian fish production from the early-1940s from the data shown in figures 1 and 3, the data are not necessarily consistent over time. Before the 1960s, it is likely that fishing was under-recorded because it was relatively unimportant to the diffuse agencies responsible for it. From the 1960s, fishing became extremely important and – as indicated in the previous section – it was recognized late in that decade that catches greatly exceeded those reported. Responsibility for Chilean fisheries also appears to have shifted over time, so it is likely that Chile's fisheries production suffers from similar enumeration problems.[77]

Nonetheless, time series data for Peruvian and Chilean fisheries from the late-1940s/early-1950s has been used for modeling purposes. Typically, the models surveyed possessed similar characteristics, employing national or regional catches for one or more species as the dependent variable. In multivariate analyses, there is a strong preference for identifying independent variables describing environmental conditions. For example, the study by Yañez, *et al.*, include indices of sea surface temperature and sea level, wind direction and stress,

---

[75]Hammergren, "Peruvian Political and Administrative Responses," 323.

[76]Boerema and Gulland, "Stock Assessment of the Peruvian Anchovy," 2228.

[77]Bernal and Aliaga note that "forerunner agencies for forestry, fish and game, were established at the end of the last [nineteenth] century, as part of the Ministry of Industry and Public Works and later under the Ministry of Trade and Commerce." Responsibility shifted to the Ministry of Agriculture after World War Two, which began to monitor and assess the hake fisheries, by far the most important before the development of the fishmeal business. The Pinochet regime overhauled the management of fisheries with the creation of the Subsecretaria de Pesca (SUBPESCA) in 1976 and the Servicio Nacional de Pesca (SERNAP) in 1978. P.A. Bernal and B. Aliaga, "ITQs in Chilean Fisheries," in A. Hatcher and K. Robinson (eds.), *The Definition and Allocation of Use Rights in European Fisheries: Proceedings of the Second Concerted Action Workshop on Economics and the Common Fisheries Policy, Brest, France, 5-7 May 1999* (Portsmouth, 1999), 117.

upwelling, and an aggregate Southern Oscillation Index among the independent variables.[78] While these variables are shown to be statistically significant, making their inclusion wholly appropriate, there should be cause for concern that selecting independent variables from an environmental "menu" limits the range of possible hypotheses and conclusions that may be derived from the exercise. In particular, hypotheses addressing the possible effects of human intervention are less often chosen. This often *a priori* exclusion of socio-economic variables also raises an identification problem when catches are an independent variable. While Klyashtorin's suggestion that catches reflect real changes in population size is reasonable,[79] catches also reflect the resources allocated to fishing as well as market prices, fishing costs, changes in technology, and fishing legislation and management strategies. There are many reasons why fishing effort – and hence catches – may fluctuate year to year independently of population, especially in developing countries that have experienced economic and political volatility. For example, Vondruska's study clearly demonstrates the negative effect of a change in the relative price of a close substitute (soybean meal) upon the demand for fishmeal, which ordinarily would be expected to stimulate adjustments in fishing intensity.[80] Further, the 1968 and 1973 military coups in Peru and Chile, respectively, both had a long-run impact on how the fishing industry operated.[81]

It would appear then that there is scope for further time series studies of the Humboldt Current fisheries, principally to address the causal factors underpinning changes in resource abundance and to give weights to the circumstances affecting fishing industries and communities. Ideally, future studies should employ a multi-disciplinary approach so as to avoid excluding potentially interesting hypotheses. To facilitate any subsequent meta analysis, there are advantages in coordinating methodologies with studies of pelagic fisheries in other upwelling systems, such as the Benguela and California currents, to which the Humboldt current fisheries are linked by capital flows and technology transfers. Although many historians generally hold reservations about quantitative time-ordered analyses of historical processes, characterizing them as "ahistorical," they can arguably play an important role in directing such research,

---

[78]Yáñez, *et al.*, "Long-term Environmental Variability."

[79]Klyashtorin, "Long-term Climate Change," 117.

[80]J. Vondruska, "Postwar Production, Consumption, and Prices of Fish Meal," in Glantz and Thompson (eds.), *Resource Management*, 285-316.

[81]This point is emphasized in Ibarra, Reid and Thorpe, "Political Economy of Marine Fisheries Development."

especially with respect to data integrity.[82] While national trends may be reasonably well known, data relating to regional and local circumstances is less accessible. Qualitative information from documentary and archival sources regarding the perceived significance of various causal factors in the conduct of the fisheries will not only help specify statistical models, but also assist in their interpretation.

A more ambitious goal is to use historical information to draw inferences about the structure of Humboldt Current's marine animal populations in the past. This goal can only be realized using historically and ecologically based simulation modeling. Such models substitute carefully chosen assumptions for data where there is limited or no data, and thereby allow us to ask questions that cannot be answered by reference to actual data. Although simulation has hitherto played a limited role in historical research, there is potential for development of this approach as an historical tool.[83] An interesting example is Christensen's study of the Straight of Georgia ecosystem.[84] This simulation study offered a detailed picture of the likely structure of that ecosystem one hundred and five hundred years ago. There appear to be good prospects for similar historical models of the Humboldt Current ecosystem, especially since the structure of the ecosystem has previously been described in formal and empirical models.[85] With respect to parameterizing such a model, evidence collected by the Guano Administration has been widely used to estimate guano bird populations, while there is also the time series evidence on fish harvests from the mid-twentieth century discussed above.[86] With respect to climatic conditions, accumulated evidence from such diverse sources as ship's logs and rainfall records offers a chronology of El Niño events

---

[82]This problem is explained in L.W. Isaac and L.J. Griffin, "Ahistoricism in Time-series Analysis of Historical Process: Critique, Redirection, and Illustrations from US Labor History," *American Sociological Review*, LIV (1989), 873-890.

[83]D.N. McCloskey, *Econometric History* (Basingstoke, 1987), 56-59.

[84]V. Christensen, "Ecosystems of the Past: How Can We Know Since We Weren't There" (unpublished paper presented to the History of Marine Animal Populations Workshop, Esbjerg, 2000).

[85]See, for example, A. Jarre, P. Muck and D. Pauly, "Two Approaches For Modeling Fish Stock Interactions in the Peruvian Upwelling Ecosystem," *ICES Marine Science Symposium*, CXCIII (1991), 178-184; and V.F. Krapivin "The Estimation of the Peruvian Current Ecosystem by a Mathematical Model of Biosphere," *Ecological Modelling*, XCI (1996), 1-14.

[86]A number of the main studies published since the Guano Administration's establishment are cited in Duffy, "Environmental Uncertainty and Commercial Fishing."

stretching back over four and a half centuries.[87] Paleoecological evidence is available to both inform the model's initial values and indicate long-run changes in climatic conditions and fish population "regimes."[88]

What role might historical research play in such an exercise, and what benefits might it offer historians? Christensen emphasizes that the widest possible range of historical information is desirable to construct, "tune" and validate models.[89] Importantly, this analytical method can use isolated observations that under other circumstances could not be employed for formal modelling, so that evidence can be drawn from documentary evidence (official papers, newspaper reports, ship's logs, maps, diaries, etc.), traditional knowledge and oral evidence, archeological evidence and surviving material culture. Hence, identifying and evaluating relevant historical information is a prerequisite for any modeling exercise. This is likely to prove a substantial task in Latin American states, given the limited research undertaken thus far. But the benefits of such a program for historians are clear. The scope of archival research and oral history projects will inevitably extend beyond that envisaged for modelling purposes, providing a basis for subsequent analysis. Participation in the modelling process itself can establish a virtuous circle by guiding historians to identify fresh evidence and develop new hypotheses, which can be tested by re-specifying and reexamining the model.

Time series and simulation modelling strategies lie outside the mainstream of historical research. Both offer the possibility of addressing questions of interest to historians that they cannot answer using orthodox methods, and therefore deserve serious attention and support. These techniques ask a specific range of questions that will certainly not wholly satisfy the historical community addressing the wider realms of maritime activity in Latin America. However, these methods are a complement to conventional historical research, not a substitute for it.

Conclusion

This paper considers the historical development of marine fisheries in Latin America as represented in a range of published works as an initial effort to address the historical exploitation of living marine resources in the region.

---

[87]W. Quinn, V.T. Neal and S.E.A. Mayolo, "El Niño Occurrences over the Past Four and a Half Centuries," *Journal of Geophysical Research*, XCII (1987), 1449-1461, cited in Glantz, *Currents of Change*, 110.

[88]H.F. Diaz and V. Markgraf (eds.), *El Niño: Historical and Paleoclimatic Aspects of the Southern Oscillation* (Cambridge, 1992), cited in Glantz, *Currents of Change*, 112.

[89]Christensen, "Ecosystems of the Past."

Although not comprehensive, it hopefully illuminates our current understanding of the subject and the potential for its advancement in the future.

Until recently, fishing has been a marginal activity in Latin America – a region dominated by agriculture and the extractive industries. At present, there is no satisfactory explanation for the underdevelopment of the sector in the region. Nor is it entirely clear why fishing has attracted relatively little commentary from historians. The subject is mostly addressed at a highly aggregate level, and there is a great need for more detailed national studies of the development of industries and communities depending upon living marine resources, especially since it is well understood that fishing's local and regional importance greatly exceeds its contribution to national income and employment. Historians will have to account for this significance as the coastal zone and marine resources become increasingly important in Latin American economy and society.

While the fisheries development experiences of Peru, and to a lesser extent Chile, are widely known, they are somewhat isolated examples. It is also necessary to recognize that these industrial fishing industries are exceptional and do not adequately represent the past and present participation in the region's fisheries. The uniqueness of the Humboldt Current fisheries in the region is less problematic in the context of the History of Marine Animal Populations initiative. Here, the reference points are to developments in other upwelling ecosystems, principally the Californian and Benguela Currents, and the main focus is upon resources within the ecosystem under different regimes of exploitation.

This paper suggests that there may be possibilities for involving historical researchers in further studies of the Humboldt Current resources using time series and simulation modeling techniques to evaluate the status of the ecosystem before large-scale harvesting. While recognizing that modeling and conventional historical research have different objectives, there are significant benefits from close collaboration between the disciplines. Historical research is clearly the less developed half of this partnership, and will require time and resources to comprehensively review and assess available historical evidence, and to establish projects such as oral histories. While recognizing the specific goals of the History of Marine Animal Populations program, there are undoubtedly wider benefits from understanding the historical development of these hugely important fisheries, and possibly those throughout the region as a whole.

The South African Fisheries:
A Preliminary Survey of Historical Sources

Lance van Sittert

Abstract

Commercial exploitation of marine resources in southern African historically has concentrated on the west coast of the continent, where the Benguela upwelling current produces a super-abundance of particular species.[1] Over the past 200 years, whales (from c. 1785), snoek (from c. 1840), guano (from c. 1845), rock lobster (from c. 1890), hake (from c. 1940) and pilchard (from c. 1945) have fuelled successive boom-bust cycles along the coast between the Cape Peninsula and the mouth of the Kunene River. Conversely, the south and east coasts, where the warm Agulhas Current nurtures a greater diversity of marine species, but at much lower population densities, sustained only subsistence and small-scale artisanal fisheries until the second half of the twentieth century. This essay is concerned with historical records relating to the exploitation of the marine resources of the west coast/Benguela, reflecting the greater commercial importance of the west over the south and east coast fisheries. It is intended as a survey of published work that utilises historical material (defined for present purposes as pre-1945 data series). It should also be noted that, although not discussed here, the subsistence/artisanal fisheries of the Natal coast are very well documented for the period 1897-1929.[2]

---

[1]See L.V. Shannon, "The Benguela Ecosystem. 1. Evolution of the Benguela, Physical Features and Processes," *Oceanography and Marine Biology*, XXIII (1985), 105-182; P. Chapman and L.V. Shannon, "The Benguela Ecosystem. 2. Chemistry and Related Processes," *ibid.*, 183-225; L.V. Shannon and S.C. Pillar, "The Benguela Ecosystem. 3. Plankton," *ibid.*, XXIV (1986), 65-170; and R.J.M. Crawford, *et al.*, "The Benguela Ecosystem. 4. The Major Fish and Invertebrate Resources," *ibid.*, XXV (1987), 353-505.

[2]Natal Colony, "Reports of the Secretary of the Natal Harbour Department, 1897-1903; "Natal Fisheries Inspector Annual Report, 1903-1911; and "Natal Fisheries Department Reports," 1912-1929.

Generation and Context of Historical Records

The systematic collection of statistics was a comparatively late development in southern Africa. The first population census was only conducted in 1865, while the initil annual agricultural census was completed in 1888 and the first annual industrial census took place in 1916. Systematic marine research in the region was initiated by the Cape colonial state in 1896 with the appointment of a marine biologist, but this position was abolished due to budget constraints in 1906. The Cape provincial administration, charged with fisheries management at Union, revived the marine biologist's post in 1911 and "shared" the incumbent with the national Department of Mines and Industries – which initiated its own marine biological survey in 1920 – until his death in 1926. Thereafter, official marine research was the preserve of the national government, which took over administrative responsibility for the economically-important Cape marine fisheries in 1936-1940 and established a Division of Fisheries under the Department of Commerce and Industries in 1937.[3] The latter remains the state's marine research arm today, despite a bewildering array of name and departmental affiliation changes over the intervening sixty years. Academic marine research is very much a post-1945 development and, like state research with which it co-operates closely, is headquartered in Cape Town.

In this context, it is unsurprising that annual fisheries statistics series only exist from the early 1950s and in the main comprise catch figures with only patchy accompanying data on vessels, trips, etc., that make the calculation of catch per unit of effort (CPUE) possible. This reflects both the weak institutionalisation of the state (colonial, provincial or national) on the maritime commons and the low budgetary priority accorded fisheries in an economy dominated by mineral extraction and agriculture until the mid-twentieth century. The Second World War belatedly catalysed a secondary industrial revolution, giving the west coast fisheries new national significance as a source of raw material for factory production. It was in this context that statistics collection improved markedly. Because these data were also used to calibrate new conservation measures (catch quotas, closed seasons, etc.), their reliability as a reflection of the actual annual catches is doubtful, given industry opposition to such restrictions, endemic "poaching" in the high-value inshore sectors (rock lobster and abalone) and widespread recreational use of inshore species.

For the nineteenth-century colonial period, the only published marine statistical series date from the 1890s and relate to inshore fisheries (1897-1906) and guano (1871-1878 and 1897-1909). The former was compiled at the behest of

---

[3]L. Van Sittert, "The Handmaiden of Industry: Marine Science & Fisheries Development in South Africa, 1890-1939," *Studies in the History and Philosophy of Science*, XXVI (1995).

the newly-appointed marine biologist by honorary fishery officers based at most outstations. The coverage was uneven, the figures unreliable and the system severely disrupted by the South African War (1899-1902). It was discontinued by the cash-strapped colonial state shortly thereafter. The guano series, on the other hand, is both longer and more reliable, owing to the colonial state's direct pecuniary interest in this marine industry. The first data set (1871-1878) was compiled for a parliamentary inquiry into the leasing of guano islands, while figures for 1897-1909 were assembled by the government guano agent following the islands' gradual re-absorption by the state in the 1890s (see figure 1). The guano islands' series includes data for penguin eggs gathered and seal skins harvested.

Figure 1
Cape of Good Hope, 1860-1885

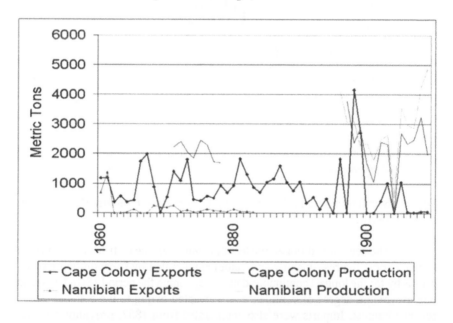

Source: *Blue Book* (Cape Town, 1879); *Returns Showing the Amount of Revenue Derived from Guano Islands of the Colony [A24-79]* (Cape Town, 1886-1909); *Statistical Register* (Cape Town, 1890-1909); *Report of the Government Guano Islands* (Cape Town, 1987-1909).

Among the other sources available for the nineteenth century is a rather idiosyncratic series spanning the period 1859-1875. This was extracted from the reports of resident magistrates/civil commissioners for the various administrative divisions of the colony published in the annual Blue Books under the heading

"Manufactories, Mines & Fisheries." These record the quantity and/or value of the catch at the various inshore outstations in each division but are predictably uneven, incomplete and unreliable. Dependent on second- or even third-hand information, this series is marred by frequent omissions and repetitions, as information from the *backveld* failed to reach Cape Town before printing deadlines (see figure 2).

Figure 2
Cape of Good Hope, 1840-1885

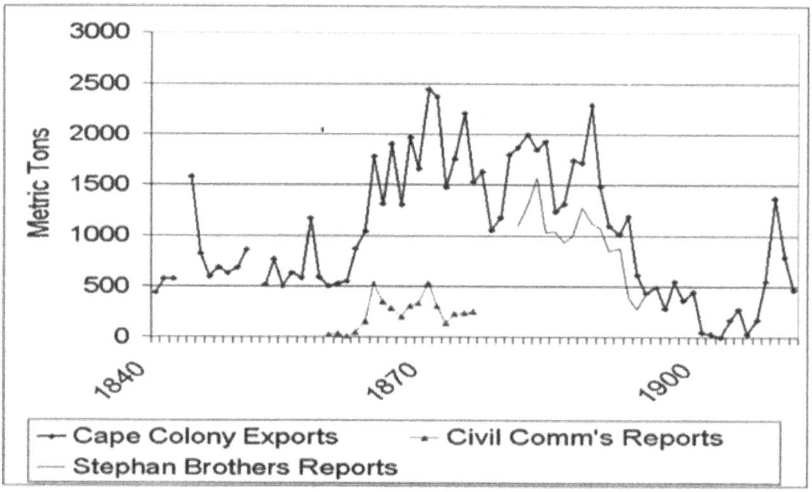

Source: See figure 1; and *Report of the Government Biologist* (Cape Town, 1896).

The second additional nineteenth-century source of fisheries data is the annual export figures (quantity and value) enumerated by Customs and Excise. This is by far the longest continuous data set, spanning the period 1840-1910, and one in which a degree of confidence is warranted owing to the colonial state's revenue interest. Imports were also enumerated from 1857, providing a useful record of fish, guano and seal production by Cape merchants on the Namibian coast – classed as imports to the Cape Colony. The Namibian data were interrupted by German annexation of the area in 1884, but resumed in 1920 following the South African conquest of the colony during the First World War.

The single biggest fishery in the colony was for snoek which, in addition to being a domestic staple for the urban poor and rural farm labourers, was dried for export to Mauritius and other Indian Ocean locales where indentured Indian labour replaced slaves on sugar plantations. The burgeoning export trade was faithfully recorded by annual Customs and Excise officials (see figure 3), as was seal harvesting (figure 4) and, from 1891 onwards, the first use of fish (rock lobster) as industrial raw material (figure 5). This series also contains figures on whalebone, whale oil, seal oil and fish oil exports/imports, but these are less useful as a gauge of the degree of exploitation of these species, being almost impossible to convert into live weight with any degree of accuracy. In the case of cetaceans, catch and population estimates have been more successfully reconstructed from other historical records.[4]

Figure 3
Cape of Good Hope, 1840-1885

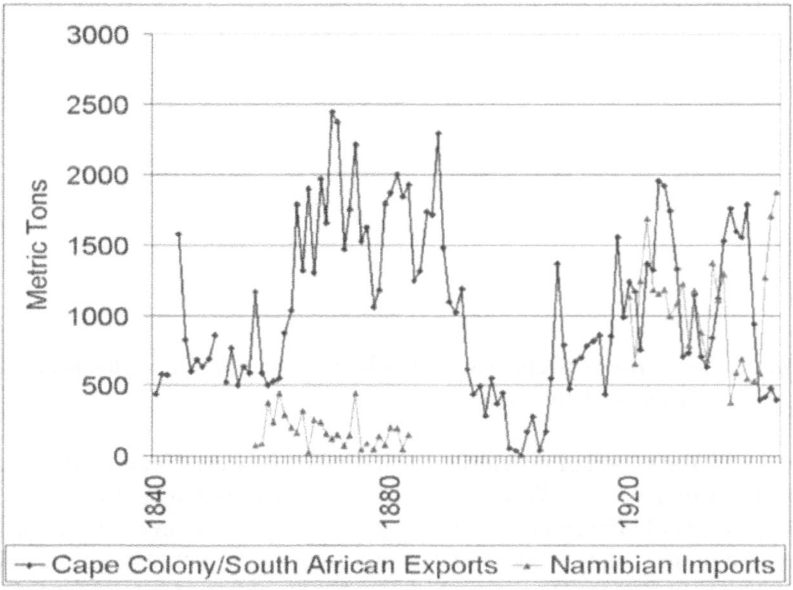

Source: *Blue Books* (Cape Town, 1886-1909). *Statistical Registers* (Cape Town, 1910-1945; *Annual Statement of Trade and Shipping* (Pretoria, various years).

---

[4]P.B. Best and G.J.B. Ross, "Catches of Right Whales from Shore-Based Establishments in Southern Africa, 1792-1975," *Report of the International Whaling Commission*, Special Issue 10 (1986), 275-289; and R. Richards and T. Du Pasquier, "Bay Whaling off Southern Africa, c. 1785-1805," *South African Journal of Marine Science* (hereafter *SAJMS*), VIII (1989), 231-250.

Figure 4
Cape of Good Hope, 1840-1885

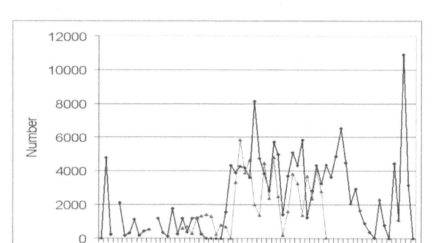

Source: *Blue Books* (Cape Town, 1886-1905); and *Statistical Registers* (Cape Town, various years).

    The period of provincial administration between Union (1910) and the outbreak of the Second World War (1940) offers a dearth of published data. The Department of Mines and Industries' marine biological survey published detailed reports of the stations worked while it prospected for new trawling grounds after 1920, and in 1935 the Division of Fisheries began publishing inshore catch figures for the Cape Town docks in its annual reports. The provincial state archives for this period, however, are rich in unpublished information, albeit unsystematic, incomplete and uncollated, recording not only catches but also boats, gear and labour employed. The one exception to this is the rock lobster canning industry, for which a monthly catch series exists for each factory from 1915 to 1931. The reason for this unusual diligence was a combination of conservation and fiscal motives, as the provincial authorities attempted to curb the industry's rapacious denudation of the fishing grounds while milking it for revenue via a controversial profits tax.[5] For precisely these reasons, however, these data must be used with

---

[5]L. Van Sittert, "'More in the Breach than in the Observance:' Crayfish, Conservation & Capitalism," *Environmental History Review*, XVII, No. 4 (1993).

caution since they not only exclude the large informal fishery for domestic consumption but were compiled from the reports of factory managers who had every incentive to under-report the quantities of fish processed to the authorities.

Figure 5
Cape of Good Hope, 1891-1895

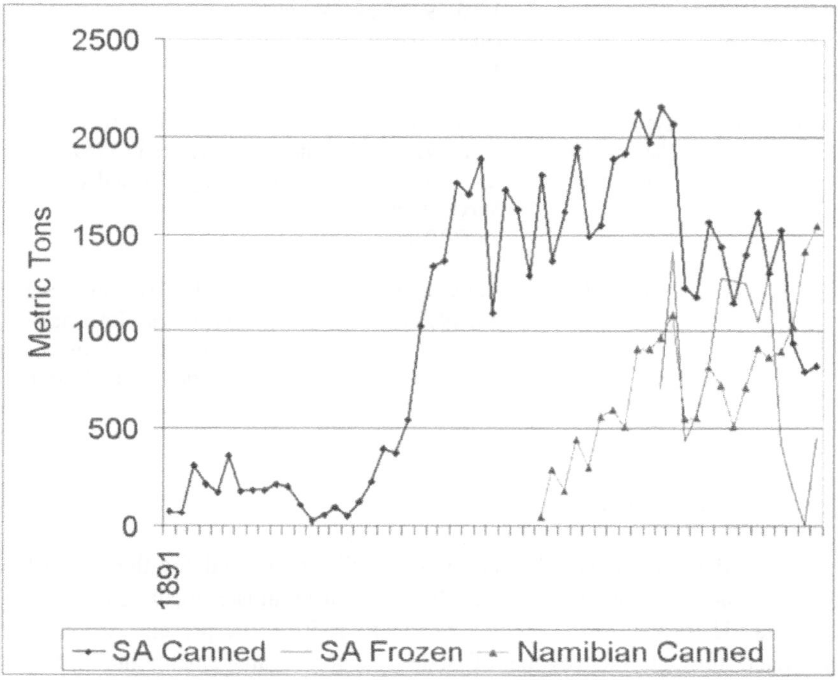

Source: *Statistical Registers* (Cape Town, 1910-1945); and *Annual Statement of Trade and Shipping* (Pretoria, various years).

The paltry catch data can be filled out through the use of national export figures published in the *Annual Statements of the Trade and Shipping of the Union of South Africa*. These are particularly useful for the rock lobster canning and freezing industry (under its own customs head from 1917) and the dried fish export trade, both of which were localised on the west coast and, in the case of the former, geared primarily to export. The industrial censuses, which commenced in 1916, are far less useful, as they only began recording catch data (in the form of the quantity of fish used as raw material) from 1946-1947 and, even then, only inventory the secondary production industry, not the fishery.

Table 1
Fishery Authority & Statistical Sources (Benguela Ecosystem)

Period	Authority	Published Reports	Statistical Series
1814-1910	Cape Colony	Marine Biologist, 1896-1906 Marine Investigations in SA, 1905-1908 Government Guano Islands, 1897-1909	1897-1906 (inshore fish) 1871-1878 + 1897-1909 (guano)
1910-1940	Cape Province National Government	Marine Biological Report, 1913-1918 Fisheries & Marine Biological Survey Report 1920-1936	1915-1931 (rock lobster) *1935-1949 (Cape Town docks)
1940-	National Government	Division of Sea Fisheries, 1937 (	1950 (Division of Sea Fisheries', Investigational Reports and *South African Journal of Marine Science*)

Note:　　* = Unpublished

All the commercial fisheries are well enumerated for the post-1945 period by the central state department charged with administering the Cape marine fisheries. These data series, which run from 1950 to the present, are chiefly generated through the regime of state resource management and industry regulation instituted in 1940-1944. They are published in annual reports, investigational reports and, since 1983, the *South African Journal of Marine Science*. The usual caveats apply, however, regarding the reliability of any and all data sets generated wholly or in part by the industry, and the complete omission from these series until recently of the burgeoning recreational/informal catch.[6]

The files of the Division of Sea Fisheries, which for the period up to 1970 have been transferred to the national archives system, also contain a substantial amount of unpublished statistical material. The same is presumably true for the South West Africa (now Namibia) Marine Research Laboratory

---

[6]A.C. Cockcroft and A.J. MacKenzie, "The Recreational Fishery for West Coast Rock Lobster, *Jasus lalandii*, in South Africa," *SAJMS*, XVIII (1997), 75-84; S.J. Lamberth, *et al.*, "The Status of the South African Beach-Seine and Gill-Net Fisheries," *ibid.*, 195-202; and B.Q. Mann, *et al.*, "An Evaluation of Participation in and Management of the South African Spearfishery," *ibid.*, 79-194.

(1959-1967), the records of which are either in the Windhoek Archives or still in departmental possession. The modern period was also marked by the generation of systematic data series on the Benguela Current ecosystem, beginning with the Pilchard Research Programme in 1951-1952 and continuing with the Benguela Ecology Programme from 1982 down to the present.

Extent and Potential of the Historical Data

It is probably fair to say that the fisheries management establishment in southern Africa, both state and academic, has a collective memory that stretches back no further than 1950. A notable exception to this generalization is the scientists working on marine mammal (whale and seal) and seabird populations, many of whom have made exemplary use of published and archival records in reconstructing historical population levels for as far back as the last quarter of the eighteenth century.[7]

The historical amnesia of fisheries managers reflects the paucity of data sets prior to 1945. Where such time series have been excavated from the archival record, as with the 1915-1931 rock lobster catch figures, these have been factored into current resource models, but their weak reliability has made them of antiquarian rather than policy interest.

In the absence of existing historical catch series, scientists and managers have tended to underestimate the extent of past exploitation and to explain the current resource crises in inshore fisheries (especially rock lobster and pelagic) in terms of the cyclical nature of the Benguela upwelling regime pulsed by the El

---

[7]See P.B. Best, "Exploitation and Recovery of Right Whales *Eubalaena australis* off the Cape Province," Division of Sea Fisheries, Investigational Report No. 80 (1970); Best, "A Review of the Catch Statistics for Modern Whaling in Southern Africa, 1908-1930," *Report of the International Whaling Commission*, XLIV (1994), 467-485; Best and G.J.B. Ross, "Catches of Right Whales from Shore-Based Establishments in Southern Africa, 1792-1975," *Report of the International Whaling Commission*, Special Issue 10 (1986), 275-289; Best and Ross, "Whale Observations from the Knysna Heads, 1903-1906," *SAJMS,* XVII (1996), 305-308; R.J.M. Crawford, *et al.*, "Trends of African Penguin *Spheniscus demersus* Populations in the 20th Century," *SAJMS*, XVI (1995), 101-118; R.W. Rand, "The Cape Fur-Seal *Arctocephalus pusillus*. 4. Estimates of Population Size," Division of Sea Fisheries, Investigational Report No. 89 (1970); P.D. Shaughnessy, "Historical Population Levels of Seals and Seabirds on Islands off Southern Africa with special reference to Seal Island, False Bay," Sea Fisheries Research Institute, Investigational Report No. 127 (1984); and P.A. Shelton, *et al.*, "Distribution, Population Size and Conservation of the Jackass Penguin *Spheniscus demersus*," *SAJMS*, II (1984), 217-258.

Niño Southern Oscillation (ENSO).[8] This simultaneously ignores the extent to which the natural environment has been altered over centuries rather than decades by human exploitation and underestimates the extent of current exploitation pressure (through a dependence on the questionable official catch figures and ignorance of informal/recreation usage).

The careful reconstruction and integration of historical catch data from secondary and archival sources into the current debate would arguably both strengthen and nuance the environmental explanations presently on offer. It could be used to test the cyclical hypothesis for the nineteenth century, cross-referenced with qualitative archival data and terrestrial data sets (for instance, rainfall records). But more important, the nineteenth-century material would also call into question the implicit assumption of a pristine marine environment prior to the mid-twentieth century that informs current models and suggest new lines of inquiry to delineate more clearly the impact of human harvesting on the Benguela ecosystem.

Two examples of the latter will suffice here. During the nineteenth century, the snoek fishery was year-round, in stark contrast to the three-month (March-May) season accepted as "natural" today. The fishing effort, moreover, shifted steadily north up the west coast through the late nineteenth century to sustain this twelve-month season. A similar northward shift in fishing effort to sustain flagging catches characterised both the rock lobster canning industry during the interwar period and the post-1945 pelagic canning and by-products industries. Similarly, the super-abundance of rock lobster in Table Bay that launched the canning industry in the last quarter of the nineteenth century has always been cast as a natural phenomenon. Yet the frequently observed massing of lobster at the sewerage outfalls in the Bay suggests that Cape Town's untreated effluent may have enhanced the effect of summer upwelling in fertilising the inshore waters and sustaining the now seemingly mythical lobster population levels reported by contemporaries.

In addition, the effect of the historical removal of whales and seals and the gathering of seabird eggs on the dynamics of the Benguela ecosystem are other avenues of enquiry worthy of investigation within an integrated framework of analysis that seeks to illuminate linkages between species and events rather than trying to understand them in isolation. To realise fully the potential of historical fisheries data, however, requires the careful reconstruction of statistics from numerous sources. This, in turn, entails an assessment of the consistency and

---

[8]D.E. Pollock and L.V. Shannon, "Response of Rock Lobster Populations in the Benguela Ecosystem to Environmental Change – a Hypothesis," *SAJMS*, V (1987), 887-899; and D.E. Pollock, *et al.*, "A Note on Reduced Rock Lobster Growth Rates and Related Environmental Anomalies in the Southern Benguela ecosystem, 1988-1995," *SAMJS*, XVIII (1997), 287-294.

integrity of the historical records before considering how they might be interpreted.

Methodological Problems

The task of reconstructing catch and effort statistics requires the development of clear hypotheses to guide data selection and the establishment of agreed standard units of measurement to facilitate both the assembly of data sets from various sources and intra- and inter-set comparability. Several archaic imperial measures were employed by the nineteenth-century Customs and Excise department, including pound, gallon, long and short ton, hundredweight and even barrel. Conversely, many observers recorded catches in number of individuals caught (presumably estimated when catches ran into millions) rather than by mass. Nor was there any consistency in the unit of measure used. Thus, for example, customs enumerators switched in 1905 from recording seal skin exports by number to weight in pounds.

A further problem is the reconstruction of catches from export figures, a calculation that involves the conversion of processed mass back into live weight. Nineteenth-century snoek exports, for example, are enumerated in pounds (lbs.) dry weight. Conversion using a standard derived from modern catches ignores the year-round nature of the nineteenth-century fishery, the well-documented seasonal change in body mass of the fish over the year and the existence of *sjinees* (Chinese) or *Bokwa* snoek – shoals of monster fish long extinct from the modern fishery.

The problems of converting processed rock lobster exports back into live weight are even more intractable, as only the crustacean's tail was canned (two to a tin), and part of the recorded weight comprised the tin and brine in which the tails were cooked. Here, contemporary norms may be a more useful guide for conversion, as both fishermen and consumers had a notorious preference for small rock lobster (less than the legislated minimum size), the former because they were paid per fish and the latter for the supposedly more succulent and tasty flesh.

The problem of data integrity has already been alluded to in relation to existing sets. In order to assess the integrity of data it is important to know the context in which they were assembled (enumerator, purpose, source of information, etc.) so as to be able to gauge its reliability with any degree of accuracy. In the case of official statistics, the general rule would seem to be that the greater the state's revenue or conservation interest, the more complete the data set but, paradoxically, the higher the likely incidence of under-reporting by producers. This problem is exacerbated in southern African by the historically low levels of fish consumption in the region and the state's corresponding lack of interest in taxing or controlling "informal" production for domestic consumption by the artisanal inshore fishery until the 1960s. Thereafter, it sought mainly to curb and suppress rather than enumerate domestic consumption of high-value export species to protect foreign exchange earnings. Non-industry production for the local (and, after 1960, increasingly black export) market and recreational use is thus a lacuna at the heart of all official fisheries data for the region.

This is not to say that informal data is necessarily non-existent or unobtainable. Angling club records, telephone surveys and oral interviews have all revealed a wealth of catch data for the modern period beyond the official statistics.[9] For the nineteenth century, too, population censuses, coastwise shipping records, newspapers and municipal market reports offer potentially fruitful sources from which the earlier dimensions of the domestic market might be reconstructed.

Figure 6
Crawfish Caught for Use of the Canning Factories of the Cape Province, 1915-1931

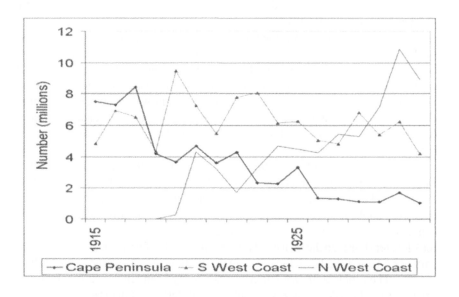

Source: Marine and Coastal Management Archive, Monthly Returns.

---

[9]B.A. Bennett, *et al.*, "Teleost Catches by Three Shore-Angling Clubs in the South-Western Cape, with an Assessment of the Effect of Restrictions Applied in 1985," *SAJMS*, XIV (1994), 11-18; Cockcroft and MacKenzie, "Recreational Fishery;" and Mann, "Evaluation of Participation."

The interpretation of historical data series requires the incorporation of ancillary qualitative data to enable human and environmental variables to be differentiated. Thus, for example, fluctuations in the official industrial rock lobster catch between 1915 and 1931 reflect price trends on the French market, not the variable availability of the resource. When the data are disaggregated by factory locale, however, the gradual northward shift in fishing effort during the 1920s – caused by the depletion of fishing grounds in the south by 1918 (see figure 6) – is clearly evident. Similarly, the marked seasonality revealed by plotting the same data by month does not chart the natural cycle of the rock lobster, but rather the constraints on fishing imposed by winter and the mandatory closed season.

Historical marine datasets thus reflect human/cultural (e.g., price fluctuations, wars, etc.) as much as natural (e.g., ENSO) variables. As marine scientists and historians are always reading these sets "against the grain," as it were (i.e., not for the purpose for which they were originally compiled), it is imperative that they are embedded as fully as possible in their original social context. Failure to do so will invariably lead to confusion or incorrect ascription of causality.

Preliminary Conclusions

Bearing in mind the dangers posed by the naive use of the historical fisheries material for the Benguela, the data, carefully embedded, have the potential to illuminate the historical development of marine animal populations in the Southeast Atlantic. It confirms, for instance, periods (cycles?) of acute fish scarcity within the Benguela system prior to the modern period and the discovery of the ENSO. The decades either side of the turn of the twentieth century, and the first half of the 1930s, were two such periods in the snoek fishery. This is reflected in the export figures and confirmed by both the St. Helena Bay data and contemporary qualitative evidence. The data sets are simply too short, however, to investigate changes in the system over the *longue durée,* while the seal and guano data are less useful because the sedentary nature of pinnipeds and seabirds encouraged more restrained or deferred use of these marine resources. In other words, the unregulated nineteenth-century fishery is the nearest thing that we have historically to a sampling of the higher trophic levels of the system.

Given the paucity of existing historical data sets for the Southeast Atlantic and the difficulties attendant on reconstructing others outlined above, it would seem that a two-tier strategy of micro- and macro-analysis would be the most fruitful way forward in working with this material. Micro-analysis would involve the identification of a few appropriately spaced sites along the western seaboard of the subcontinent (a preliminary list might be Cape Town, St. Helena Bay and Luderitz or Walvis Bay in Namibia) and the compilation and contextualisation of marine data series for each locale. The latter should integrate the widest possible range of ancillary quantitative and qualitative data, such as coastwise shipping logs, rainfall records, and magistrates' reports, to confirm integrity and enhance interpretation.

The micro-studies would then serve as controls for macro-analysis at both the national and international levels. The trends revealed by compilation of, for example, nineteenth-century dried fish export data could then be cross-checked against the trends in the Cape Town, St. Helena Bay and Luderitz/Walvis Bay series (see, for example, figure 7). Provided units of measurement were standardised, the latter could also be compared trans-nationally with micro-analysis sites from similar marine environments on the west coasts of South America (Peru/Chile), North America (California) and Australia (Western Australia).

Figure 7

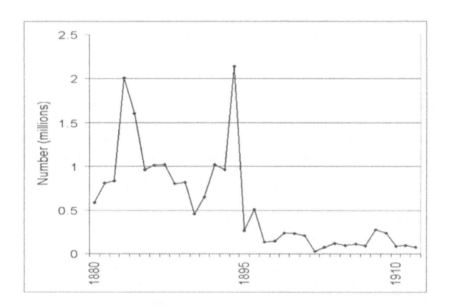

Source: Cape Archives: PAN69, K59/3, Stephan Brothers to the Administrator, 23 August 1913.

The Potential for Historical Studies of Fisheries in Australia and New Zealand

Malcolm Tull and Tom Polacheck

Abstract

This paper presents a preliminary appraisal of primary sources that might prove valuable in investigating the impact of human harvesting on fish populations off Australia and New Zealand. Since for most of their history the Australasian fisheries have been dominated by small family operations, few company records exist. This means that the major sources have been generated by government fisheries departments and other agencies, as well as by reports and proceedings of government inquiries. The paper focuses on printed statistical sources because further research is required to locate archival and other sources, especially in New Zealand. Secondary literature is also used to present an historical snapshot of Australian and New Zealand fisheries. The paper begins with an outline of the development of commercial fishing in both countries; then discusses statistical sources; and finally assesses their potential for the HMAP project. A case study of the Southeast Australian Trawl Fishery is also included; its main finding is that this important multi-species fishery will yield much evidence to test the hypotheses that lie at the core of the History of Marine Animal Populations project.

Introduction

In 1988, Giulio Pontecorvo made what he termed "a modest proposal" for an improved global fishery database to be administered by the Food and Agriculture Organisation (FAO) which would provide more complete information on the health of world fisheries.[1] In particular, he called for more comprehensive time series on capital, labour, prices and stock assessments. This led to a debate about the costs and benefits of data collection and the need to ensure that data are collected "not just for the benefit of a few ivory tower academics."[2] Naturally

---

[1]G. Pontecorvo, "The State of Worldwide Fishery Statistics: A Modest Proposal," *Marine Resource Economics*, V, No. 2 (1988), 79-81.

[2]J.A. Gulland, "Comments on Giulio Pontecorvo's 'The State of Worldwide Fishery Statistics: A Modest Proposal,'" *Marine Resource Economics*, VI, No. 1 (1989), 85-86. See also M.A. Robinson and F.T. Christy, Jr., "Comment on Professor

enough, fisheries managers tend to have a narrower focus than scientists, economists and other scholars interested in marine animal populations. For example, in 1996, a review of the Australian Fisheries Management Authority (AFMA) concluded that there was insufficient information on stock levels, fishing effort and catch statistics to enable an accurate assessment of the Authority's effectiveness.[3] The AFMA responded as follows:

> In reality fisheries managers do not have a great need for
> statistics. A manager will learn much more developments in a
> fishery and what management initiatives are appropriate by
> maintaining good industry contact than from statistics. A lot of
> the hype about the need for statistics comes from researchers
> who are not provided with the where-with-all to collect data
> specifically tailored to their particular needs. Instead they look
> to fishers and fisheries managers to supply them with data that
> the researchers themselves consider inadequate.[4]

This rather narrow view fails to recognize that apart from management decision-making, fisheries statistics are used for a variety of purposes, including international reporting and policy formation, and scientific studies on ecosystem dynamics. It can be argued that decisions by a wide range of actors – ranging from the World Bank to small non-government aid agencies – can exert more influence on the health of global fish stocks than fisheries managers, and that these bodies need access to comprehensive data on world fisheries.[5] Thus, while it is legitimate to question the cost and benefits of data collection, there is a large and growing market for worldwide fisheries statistics.

In this respect, projects such as the History of Animal Population (HMAP) project can play a valuable role in fostering fisheries research. The aim of HMAP is to integrate marine and aquatic ecology, history and palaeoecology to improve understanding of the role of marine resources in human history, to investigate long-term changes in marine animal stocks and to examine the

---

Pontecorvo's 'The State of Worldwide Fishery Statistics: A Modest Proposal,'" *Marine Resource Economics*, VI, No. 1 (1989), 83-84.

[3]Australian National Audit Office, *Commonwealth Fisheries Management, Performance Audit Report*, No. 32 (Canberra, 1996), I, 4.

[4]*Ibid.*, II, 84.

[5]Gulland, "Comments," 85.

ecological implications of these changes.[6] A key outcome of the project will be the creation of estimates of catches and fishing effort in the period from 1500 to 2000.

The Australian and New Zealand Fishing Industries

Australia, which was settled by the British in 1788, is an island nation with over 37,000 kilometres of coastline. New Zealand, another island nation, was settled by the British in 1840. It lies about 1600 kilometres to the west of Australia across the Tasman Sea. In the late 1970s, both countries claimed sovereignty over all marine resources within 200 nautical miles of their respective coastlines. Australia now controls an exclusive economic zone (EEZ) of about 8.94 million square kilometres, about 1.2 times its landmass. New Zealand controls an EEZ of four million square kilometres, about fifteen times its landmass. These zones are among the largest in the world and provide both nations with legal jurisdiction over vast marine resources.

Yet despite having two of the largest fishing zones on the world, both countries are only minor fishing nations by international standards. In 1998, Australian fisheries production totalled 228,983 tonnes, or 0.2 percent of the world total, while New Zealand's output accounted for 673,106 tonnes, or 0.6 percent of global production. By contrast, China, the world's largest fishing nation, had a catch of thirty-eight million tonnes, or thirty-two percent of total world production.[7] This is not simply a reflection of population size, although Australia and New Zealand's populations are only nineteen million and 3.8 million, respectively. To explain why both Australia and New Zealand are still minor fishing nations, and to provide an historical context for our discussion of sources, we offer a brief review of the history of these nations' fishing industries.

The Development of Australia's Fishing Interests

Until relatively recently, Australian historians have neglected maritime history, including the fisheries.[8] Fishermen and their communities, "both in the cities and

---

[6]P. Holm, T.D. Smith and D.J. Starkey, "History of Marine Animal Populations: Testing Ecological Hypotheses" (http://www.cmrh.dk/hmappros.html).

[7]Food and Agriculture Organization (FAO), Fishery Production by Country, 1998 tonnes (preliminary data).

[8]See F. Broeze, *Island Nation. A History of Australians and the Sea* (St. Leonards, NSW, 1998); and M. Tull, "Maritime History in Australia," in F. Broeze (ed.), *Maritime History at the Crossroads: A Critical Review of Recent Historiography* (St. John's, 1995).

dotted around the coast have never mustered the numbers to make any significant inroads into the social or economic histories of urban or rural Australia."[9] While steps have been taken to correct this deficiency, the literature on general fishing (the capture of scale fish, crustaceans and molluscs) is still relatively limited.[10] A comprehensive bibliography of primary and secondary sources on fishing history has been produced by Smith and Tull, while specialized bibliographies on whaling have been assembled by Forster.[11] Tull has published a bibliography of university theses on Australian maritime history that includes sixty-seven entries on fishing.[12]

When British settlers arrived on the coasts of Australia and New Zealand in the eighteenth and nineteenth centuries they did not encounter *terra nullius*, that is, land belonging to no one; both countries had indigenous populations. Australia's Aborigines, although not a seafaring people, had long fished the rivers, beaches and estuaries along the island's vast coastline in a sustainable manner. Aboriginal rock paintings depicting fish such as barramundi testify to the cultural importance of fish. Rock paintings also recorded the visits of Indonesian fishers with whom pearl shell, trepang and other marine products were traded.[13] In 1992, the Australian High Court recognized the existence of Aboriginal land rights, and in the following year the federal government passed

---

[9]F. Broeze, "From the Periphery to the Mainstream: The Challenge of Australia's Maritime History," *The Great Circle*, XI (1989), 2-3.

[10]For a general survey, see M. Tull, "The Development of the Australian Fishing Industry: A Preliminary Survey," *International Journal of Maritime History*, V, No. 1 (June 1993), 95-126.

[11]H. Smith and M. Tull, *The Australian Fishing Industry: A Select Historical Bibliography* (Perth, 1990); H. Forster, *The South Sea Whaler: An Annotated Bibliography of Published Historical, Literary, and Art Material relating to Whaling in the Pacific Ocean in the Nineteenth Century* (Sharon, MA, 1985); and Forster, *More South Sea Whaling. A Supplement to The South Sea Whaler: An Annotated Bibliography of Published Historical, Literary and Art Material relating to Whaling in the Pacific Ocean in the Nineteenth Century* (Canberra, 1991).

[12]M. Tull, "A Bibliography of University Theses on Australian Maritime History," *International Journal of Maritime History*, VIII, No. 1 (June 1996), 199-246. Some of the theses have substantial technical and scientific content.

[13]See N. Burningham, "Aboriginal Nautical Art: A Record of the Macassans and the Pearling Industry in Northern Australia," *The Great Circle*, XVI, No. 2 (1994), 139-151.

the Native Title Act, which granted land rights to Aboriginal people.[14] This act has been used to grant Aboriginals limited customary rights to fish for non-commercial and commercial purposes.

For a limited period in the early nineteenth century, seals and whales provided Australia's first commercial exports, with whaling remaining its major export industry until 1833, when it was overtaken by wool. But by the 1840s catches were falling. As is well known, in the absence of regulation, all fisheries tend to follow a similar historical cycle. Initially, there is rapid growth as the potential of the fishery is tapped, but this is followed by over-exploitation and then decline. Norwegian whalers, equipped with factory ships and modern harpoon guns, revived the industry at Jervis Bay in New South Wales in 1912; at Frenchman's Bay, near Albany in Western Australia, also in 1912; and at the remote West Australian outpost of Point Cloates in 1913.[15] Dwindling catches and the outbreak of the First World War, however, caused the Norwegians to leave in 1916.

An Australian firm, the North West (Aust.) Whaling Company, attempted to establish operations at Point Cloates but went bankrupt within two years. Its whaling equipment was leased to a Norwegian company that operated profitably between 1925 and 1929. Although foreign whalers began operating off the northwest coast in 1936, capturing over 7000 whales within three seasons of whaling, no further shore-based whaling was carried out until after the Second World War. By 1948 world shortages of fats had caused the price of whale oil to reach about six times the prewar price, and five companies began whaling on the Australian coast. They continued until 1963, when declining numbers of whales, and a ban on the taking of humpbacks, led to the closure of all the Australian whaling stations except the one at Albany (Cheynes Beach). This station continued hunting sperm whales until 1978, when declining returns led to its closure and the end of the Australian whaling industry. The writing was on the wall for the industry worldwide, and in 1985 the International Whaling Commission (IWC) introduced a ten-year moratorium on whaling.

In the second half of the nineteenth century, another form of fishing – pearling – became important in Western Australia, Queensland and the Northern Territory. The main objective of commercial pearling was the collection of pearl shell, rather than the pearls themselves, the latter being regarded as a welcome bonus. Pearl shell was valued for its hardness and durability and was used for a

---

[14]G.D. Meyers, "Implementing Native Title in Australia: The Implications for Living Resources Management," *University of Tasmania Law Review*, XIV, No. 1 (1995), 1-28.

[15]For details, see M. Tull, *The Development of Western Australia's Fishing Industry, 1900-1980: A Preliminary Overview* (Murdoch, WA, 1990).

wide variety of purposes including button and ornament making, as well as the manufacture of dial-plates for prismatic compasses. By 1900 the industry employed a multicultural workforce of about 4000 people and generated exports to the value of $430,000.[16] In the heyday of the industry before the First World War, Australia supplied between one-half and three-quarters of the world's output of pearl shell. But after the First World War the industry declined due to increasing competition and falling prices for pearl shell. Its fate was sealed by the development of plastics, which by the late 1940s offered a cheap and effective alternative to pearl shell in most uses. The industry was, however, saved from total extinction by the Japanese, who pioneered the technique of pearl culture. A pearl farm was established in Western Australia in 1956, and pearl culture has since developed into a valuable aquaculture industry. In 1997-1998, Australian pearl production accounted for eighteen percent of the gross value of total fisheries production of $1.86 billion.

Until the 1940s general fishing was a "Cinderella industry" of limited economic importance in Australia. That the scale of fishing activity was limited was due to a variety of factors, including consumer tastes (mainly a preference for meat rather than fish), low productivity of the oceans, and greater opportunities for wealth generation in land-based industries.[17] Since the Second World War, the exploitation of high-value crustacean, mollusc and pelagic fisheries has created a lucrative export industry. The most spectacular success story is that of the Western Rock Lobster fishery, which was transformed from a minor industry serving local markets into Australia's most valuable export fishery. The introduction of management controls in 1963 has ensured that it remains sustainable.[18] The Southern Bluefin tuna fishery was developed in the late 1960s primarily to cater for the Japanese *sashimi* market. But dwindling stocks checked expansion, and in 1984 the government introduced an individual transferable quota system to safeguard the fishery.[19] Another valuable fishery, also developed

---

[16]All financial data prior to 1966 have been converted to dollars on the basis that £1 = $2.

[17]Tull, "Development of the Australian Fishing Industry," 114-121.

[18]See M. Tull, "Profits and Lifestyle: Western Australia's Fishers," *Studies in Western Australian History*, XIII (1992), 92-111; H. Gray, "Askinnin' the Pots. A History of the Western Rock Lobster Fishery" (Unpublished PhD thesis, Murdoch University, 1999); and Gray, *The Western Rock Lobster. Panulirus Cygnus. Book 2: A History of the Fishery* (Geraldton, 1999).

[19]See S. Adams, "The International Management of Southern Bluefin Tuna: Consensus or Conflict?" (Unpublished Paper presented to Conference on "Shaping Common Futures: Case Studies of Collective Goods, Collective Actions in East and

in the 1960s for the Japanese market, was the northern prawn fishery. By 1997-1998, rock lobsters accounted for twenty-eight percent of the total value of fishery exports of $1.5 billion, followed by pearls at nineteen percent, prawns at sixteen percent and abalone at thirteen percent.

Although pearling remains the mainstay of the aquaculture industry, the farming of tuna, salmon and freshwater crayfish has expanded rapidly since the 1980s. By 1997-1998, aquaculture (including pearls) accounted for twenty-six percent of total production. But Australia has never produced enough seafood to meet domestic demand and currently imports over half of the seafood consumed, mainly prawns, frozen fish fillets and canned fish. In 1997-1998, imports totalled $818.8 million, the main suppliers being Thailand and New Zealand.

The Development of New Zealand's Fishing Interests

In New Zealand, the focus of the early settlers was on exploiting the resources of the land rather than the sea, and so "the harvesting of New Zealand's rich coastal waters was carried out on a small scale and in a haphazard fashion."[20] Although New Zealand's offshore supports over 1000 species of fish, the waters are not very productive, and only about 100 species have ever been of commercial significance. This perhaps explains the paucity of the historical literature, especially on general fishing.[21]

As in Australia, the issue of indigenous fishing rights has proved controversial in New Zealand. The country was inhabited by a Polynesian race called the Maori, who had a tradition of seafaring and maritime activity.[22] Archaeological sites between North Cape and the Bay of Plenty have produced large quantities of snapper bones and a variety of types of fishhooks.[23] The Maori caught fish with nets and lines and took eels and lampreys with pots set in

---

Southeast Asia," Asia Research Centre, Murdoch University, 7-9 October 1999).

[20]T. Brooking, "Economic Transformation," in G.W. Rice (ed.), *The Oxford History of New Zealand* (2nd ed., Oxford, 1992), 234.

[21]Canterbury University Press is planning to publish in 2002 a history of New Zealand fisheries by D. Johnson, entitled *Fish: A History of the New Zealand Fishing Industry*.

[22]For a comprehensive account of Maori fishing, see E. Best, *Fishing Methods and Devices of the Maori* (1929; reprint, Wellington, 1977).

[23]J.M. Davidson, "The Polynesian Foundation," in Rice (ed.), *Oxford History of New Zealand*, 21.

weirs.[24] Article 2 of the Treaty of Waitangi, signed with the British in 1840, ostensibly protected the Maori *taonga* (treasures), including the fisheries. In practice, however:

> The laws assumed that the Crown had an unrestricted right to dispose of the inshore and offshore fisheries; that the *Treaty of Waitangi* had no real bearing on the fisheries issue; that Maori had had no commercial component (despite substantial evidence to the contrary); and that non-Maori interests could be licensed for commercial exploitation while Maori interests could be provided for in non-commercial reserves near to their major villages.[25]

Maori claims for both customary and commercial fishing rights simmered for 150 years. Matters came to a head in 1986 when the government, in an attempt to deal with dwindling fish stocks, introduced the Quota Management System. This created a regime of property rights for commercial fishers. These rights were granted on the basis of commercial catch history, ignoring the interests of groups such as the Maori and recreational fishers. The Maori successfully argued in court that this breached their treaty rights, and after much controversy the government passed the Treaty of Waitangi Settlement Act of 1992. This guaranteed the Maori both customary and commercial fishing rights. The government agreed to fund the Maori in a 50/50 joint venture with Brierley Investments Ltd. to take over New Zealand's largest fishing company, Sealord Products Ltd. The NZ $175 million paid for the half share gave the Maori control over about one-third of commercial fishing quotas in New Zealand.[26] The Treaty of Waitangi Fisheries Commission (*Te Ohu Kai Moana*) was created to administer Maori fishing interests, though the allocation of the quota between *Iwi* (tribal groups) has led to considerable acrimony.[27]

Sealers and whalers explored the New Zealand coastline long before it became a British colony. This activity was undertaken by Americans, Britons and

---

[24]A.H. McLintock (ed.), *An Encyclopaedia of New Zealand* (Wellington, 1966), I, 679-687.

[25]A. Parsonson, "The Challenge to Mania Maori," in Rice (ed.), *Oxford History of New Zealand*, 195.

[26]*New Zealand Official Yearbook* (Wellington, 1998).

[27]Hon. D. Kidd, MP, "A Minister's Perspective on Managing New Zealand Fisheries" (Unpublished Paper presented to "Fish Rights 99" Conference, Fremantle, November 1999).

others engaged in the South Sea fisheries which, although originally concentrated in Bass Strait, extended to the seas around New Zealand. In 1840 the number of American ships engaged in whaling in New Zealand waters was estimated at between 600 and 700, employing 18,000 seamen; in 1845 their catch was estimated to be worth NZ $2.8m.[28] John Guard, an ex-sealer at Te Awaiti, Cook Strait, established the first shore-based whaling station in 1827, and whale exports quickly became important.[29] The industry peaked in the late 1830s, after which ruthless exploitation led to declining catches. As J.B. Condliffe observed, "the whaling and sealing carried on in New Zealand waters in the early nineteenth century is a classic illustration of the wanton destruction of an important national asset by unscientific and unrestrained exploitation."[30] The whaling industry continued on a much-reduced scale in the twentieth century, but due to declining yields and IWC restrictions, it ceased in the early 1960s.

As the population grew, there was increased demand for general fishing for domestic consumption and later export. In the 1880s, the government, which was keen to start new industries, introduced a bounty of two old pence for every pound of fish exported.[31] This was successful, and cod and oysters were exported to Australia. But total yearly exports were only about 600 tons, while imports averaged about 1000 tons per annum. Activity was confined mainly to inshore grounds close to population centres.[32] Before the Second World War, the main method of fishing was long-lining for groper (hapuku), ling and hake (kingfish); hand-lining for blue cod; and trawling and Danish seining for flounder, snapper, tarakihi, gurnard, john-dory and a variety of other species. Set-nets were used in bays and estuaries to capture flounder, flatfish and snapper.[33]

In the 1860s, attempts were made to establish a freshwater fishery with the introduction of British salmon and trout. The first brown trout were reared from ova imported from Tasmania in 1867. The trout acclimatized well, and by the 1880s and 1890s the rivers and lakes of New Zealand carried large stocks of

---

[28]J.B. Condliffe, *New Zealand in the Making. A Study of Economic and Social Development* (2nd ed., London, 1959), 130-131.

[29]N. Prickett, "The New Zealand Shore Whaling Industry," in S. Lawrence and M. Staniforth (eds.), *The Archaeology of Whaling in Southern Australia and New Zealand* (Gundaroo, NSW, 1998), 48-53.

[30]Condliffe, *New Zealand in the Making,* 130.

[31]G.H. Scholefield, *New Zealand in Evolution* (London, 1909), 291.

[32]*New Zealand Official Yearbook* (Wellington, 1912), 854-858.

[33]*New Zealand Official Yearbook* (Wellington, 1938), 464.

these fish. But they were mostly used for recreational fishing in what New Zealanders claimed to be "the Angler's Paradise."[34] A crayfish (rock lobster) fishery grew rapidly after 1947, when an export market for tails developed in the US. The catch peaked at 130,815 hundredweight in 1956 but has since declined. It remains, however, the most valuable species caught.

Since the creation of the EEZ in 1978, fishing output has taken off as New Zealand's fishers have exploited the deepwater stocks within their zone. Species such as hoki, hake, ling, orange roughy, oreo dories, squid and silver warehou are targeted at depths ranging from 200 to 1200 metres. This is one of the deepest fisheries in the world. Total landings of fish doubled between 1979 and 1983 to reach 282,000 tonnes, and then peaked at 503,000 tonnes in 1992.

As in Australia, aquaculture has grown in importance. The main species farmed are greenshell mussels, salmon and Pacific oysters. In 1997, exports of mussels and salmon totalled NZ $84.5 million and NZ $31.5 million, respectively. The main export markets are Australia, Japan and the US. In 1997, total exports were valued at NZ $1125.4 million and comprised mainly frozen fish, rock lobster and shellfish. Imports are mainly canned and prepared fish, such as salmon, sardines and herring.

Fishing Statistics

There are several possible measures of fish catches: numbers, weight or value. All have limitations: number data lack homogeneity and require conversion to weight on the basis of ratios; weight data overstate the importance of bulky, low-value seafood; and value data give rise to the problem of finding an adequate price deflator. No single measure is adequate, but from an economic perspective, the value of catches is probably the most important indicator, as it conditions the willingness to invest in the industry. While value data are generally available for total production and overseas trade, for the most part historical discussion of individual species will be in terms of tonnages, simply because value detail is often lacking. Australian and New Zealand statistics are now examined in turn.

Australian Statistics

In line with British practice, Australia was required to provide statistical information to the Colonial Office in London.[35] From 1822 until the granting of

---

[34]*Ibid.*, 469.

[35]See C. Forster and C. Hazlehurst, "Australian Statisticians and the Development of Official Statistics," in *Australian Yearbook* (Canberra, 1988), 1-91; and W. Vamplew (ed.), *Australian Historical Statistics* (Broadway, NSW, 1987), X.

responsible government in the mid-1850s (with the exception of Western Australia, which had to wait until 1890), each Australian colony prepared annual returns that were published in *Blue Books*. These included a variety of statistics on the progress of the colonies, including a limited amount of material on the fisheries. Following the granting of responsible government, the *Blue Books* were replaced by *Statistical Registers*. In 1891, the New South Wales Government Statistician, T.A. Coghlan, prepared a *Statistical Account of the Seven Colonies of Australasia*, which for the first time provided statistics on the six Australian colonies and New Zealand. Eleven editions were published, ending in 1903-1904.[36] The detailed statistics collected by government statisticians were "vital to an interventionist State in a new land" and were at the forefront of international best practice.[37] But the existence of six separate statistical organisations led to a lack of uniformity in data collection and publication, which has made it difficult for historians to produce meaningful statistics for Australia as a whole. Nevertheless, in 1962, N.G. Butlin produced a series of macroeconomic indicators, which still remains the starting point for long-run analysis of Australian economic growth.[38]

The colonies joined in a Federation in 1901, and the federal government created a Commonwealth Bureau of Census and Statistics (CBCS) in 1905. This led to gradual improvements in the uniformity of data collection, but there were now problems with duplication of effort. In 1956, the federal government began the integration of its statistical service with those of the states, a process completed by 1958. In 1975, the CBCS was renamed the Australian Bureau of Statistics (ABS).

The minor status of the fishing industry compared to land-based pastoral and agricultural activities meant that, with the exception of whaling and pearling, it did not receive much coverage in the official statistics until after the Second World War. The first reference to fisheries in Coghlan's pioneering *A Statistical Account of Australia and New Zealand* appeared in the 1901-1902 edition, which also briefly commented on the decline in whaling. The final edition of 1903-1904

---

[36]Due to federation, the last two editions were renamed *A Statistical Account of Australia and New Zealand*.

[37]P. Groenewegen and B. McFarlane, *A History of Australian Economic Thought* (London, 1990), 115.

[38]N.G. Butlin, *Australian Domestic Product, Investment and Foreign Borrowing, 1861-1938/39* (Cambridge, 1962). See also Butlin, "Contours of the Australian Economy, 1788-1860," *Australian Economic History Review*, XXVI, No. 2 (September 1986), 96-125.

claimed that total fisheries production in Australia and New Zealand was worth $2,156,000, although this was mainly from pearling in Australian waters.[39]

Figure 1
Australian fisheries gross output, 1861 to 1900
(1910-11=1000)

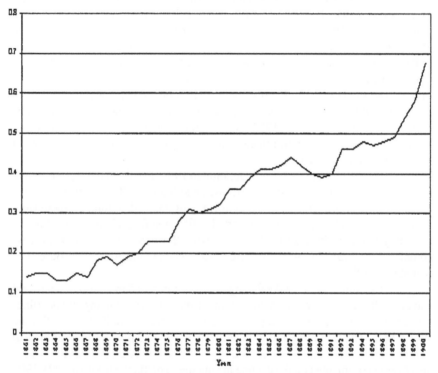

Notes: Deflated using GDP deflator 1910-1911 = 1000, from W. Vamplew
(ed.), *Australian Historical Statistics* (Broadway, NSW, 1987).

Source: N.G. Butlin, *Australian Domestic Product, Investment and Foreign
Borrowing, 1861-1938/39* (Cambridge, 1962), 133.

When calculating his pioneering estimates of economic growth, N.G.
Butlin noted that fisheries production was not consistently presented: after 1900
it included the supply of fresh and frozen fish to consumer markets, whereas

_____

[39] *A Statistical Account of Australia and New Zealand, 1903-04* (Sydney, 1904),
1010.

before 1900 it also included the processed products of the whale fisheries. The latter had become insignificant by the late 1870s. Moreover, due to data deficiencies, he estimated Australian fisheries output on the basis of data from just New South Wales and Tasmania.[40] Figure 1 illustrates Butlin's data on fisheries output between 1861 and 1900, expressed in 1910-1911 prices. The demise of whaling did not affect long-term growth, as output expanded steadily at about four percent per annum between 1861 and 1900.

Since Federation, various government bodies have collected three main types of statistical data on Australian fisheries: landing statistics collected by each state and the ABS; Commonwealth catch logbook systems, which tend to provide more detailed records of catch and effort than state data; and records of fish sales collected by central marketing authorities or management bodies.[41]

The first *Official Yearbook of the Commonwealth of Australia* (1908) noted that the coverage and quality of data collection varied between the states to such an extent that it was not possible to present nationwide statistics.[42] For example, South Australia and Tasmania were not included in the *Yearbook*'s tonnage series until 1925. Discrepancies in ratios used to convert fish numbers to weights also impaired the comparability of the data. There is the possibility that some fishers under-reported catches to reduce their tax bills. Indeed, it has been claimed that for this reason "many of them went to a great deal of trouble to ensure that there were no records of their activities."[43] Information in the *Statistical Registers* and *Yearbooks* on the number of licensed fishers provides a guide to employment, but these statistics include inactive and part-time fishers and are not a direct measure of labour effort. Until 1950, amateur fishers were included in the license data, and a number of changes were made in the 1960s that limit comparability with earlier years. Thus, at best the data in the *Statistical Registers* and *Yearbooks* provide only an approximate guide to activity trends in the industry.

Table 1 uses Butlin's estimates and the official statistics to provide an overview of the progress of the industry from 1861 to 1980. The value of the catch (1910-1911 = 1000) increased by a factor of fifty-five between 1861 and 1980, although most of the increase occurred from the 1950s onwards. This was due to the exploitation of high-value crustacean, mollusc and pelagic (mainly

---

[40]Butlin, *Australian Domestic Product,* 125 and 134-135.

[41]Australia, Bureau of Rural Resources, *Twenty-five Years of Australian Fisheries Statistics* (Canberra, 1991), 2.

[42]*Official Yearbook of the Commonwealth of Australia* (Melbourne, 1908), 387.

[43]D.H. Borchardt (ed.), *Australians: A Guide to Sources* (Sydney, 1987), 254.

tuna) fisheries. The postwar expansion of the industry attracted more entrants, but fishing has never been a major employer. In the 1961 census, only 0.2 percent of the workforce was employed in fishing. Catch per boat, measured in 1910-1911 prices, increased from $283 in 1950 to $1684 in 1980, clear evidence of an increase in fishing efficiency. This was, of course, mainly due to the technological changes that transformed postwar harvesting operations.

Table 1
Gross Output, Boats and Employment in Australia's Fisheries, 1861-1980

	Gross Catch ($ Million, 1910-1911 Prices)	Boats	Fishers
1861	0.28	NA	NA
1870	0.34	NA	NA
1880	0.64	NA	NA
1890	0.78	NA	NA
1900	1.36	2230+	6693+
1910	1.57	3787	9727
1920	1.69	4671	11911
1930	2.10	5729	12011
1939	2.23	6397	11681
1950	3.00	10587	17898
1960	4.75	9188	15,356%
1970	8.28	10695	17867
1980	15.31	9090*	21,000#

Notes: Calendar years up to 1900, financial years thereafter. Catch values deflated using GDP price deflator index in Vamplew (ed.), *Australian Historical Statistics*, 219. NA = Not available; % = the removal of amateur fishers from the totals after 1950 means that fisher numbers are not comparable with earlier years. + = based on incomplete data for 1901. * = 1985. # = estimate for 1987.

Sources: See figure 1; *Australian Yearbooks*, various years; and Australian Fisheries Service, *Background Fisheries Statistics* (Canberra, 1990).

In 1991, the Bureau of Rural Resources (BRR) produced a detailed series of fisheries production statistics by weight for 1964-1965 to 1989-1990.

The BRR attempted to produce consistent and comprehensive time series of commercial catch of live-weight marine animals but encountered serious problems with the data, most notably data entry errors; inaccurate species identification; confusion over jointly-managed fisheries (data collection procedures by the federal government and states have sometimes differed, and double-counting has occurred); misreporting of catches (due to differences in logbook design and resolution of data); catch figures that differed from live weight (on board preparation of catch method of storage and date of entry in logbooks affects the weights recorded, and different assumptions have led to incompatible state data); and inaccurate industry returns (the extent of under- or over-reporting is hard to quantify).

Inaccurate species identification is an issue that has plagued the industry for many years, reducing the accuracy of fishing statistics and leading to claims of misleading marketing. Moreover, imported fish have often been sold as local fish: for example, in Western Australia, South African fish has been sold as snapper and dhufish.[44] To reduce confusion the Australian Fisheries Service published a guide in 1985 recommending marketing names for fish.[45]

Nevertheless, the BRR has provided a valuable and comprehensive suite of data for individual species and for sales at fish markets in New South Wales, Victoria and Queensland. These data were used in a major reference work published by the BRR and the Fisheries Research and Development Corporation.[46] Apart from the various state departments of fisheries, data for the period after 1989-1990 are available from the Australian Bureau of Resource Economics (ABARE), which since 1991 has published an annual report providing comprehensive fisheries data for all states.

Figure 2 shows the growth of total landings of fish in Australia and New Zealand between 1920 and 1993, measured in thousand metric tonnes. Australia was more important until New Zealand's industry took off in the late 1970s. By the 1990s, landings of fish in New Zealand were more than double those in Australia.

---

[44]W.D. Scott and Co. Pty. Ltd., *A Report to the Honourable, the Minister for Fisheries and Fauna on a Pilot Study into the Economic Future of the Wet Fish Industry in Western Australia* (Perth, 1969), 19.

[45]Australian Fisheries Service, *Recommended Marketing Names for Fish* (2nd ed., Canberra, 1988).

[46]P.J. Kailola, *et al.*, *Australian Fisheries Resources* (Canberra, 1993).

Figure 2
Landings of Fish, Australia and New Zealand, 1920-1993

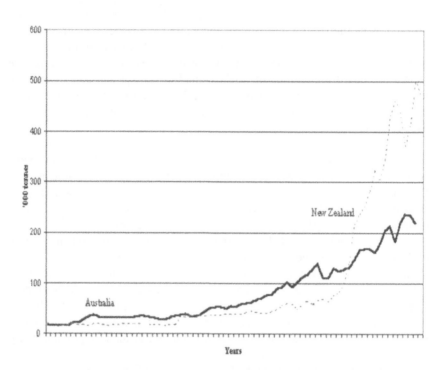

Source: B.R. Mitchell, *International Historical Statistics. Africa, Asia and Oceania 1750-1993* (3rd ed., London, 1998), 330.

The Southeast Australian Trawl Fishery[47]

A trawl fishery developed off the southeast coast of Australia in the early 1900s. Its location is illustrated in figure 3. This quickly developed into the dominant Australian finfish fishery and has been the primary supplier of fresh fish for the domestic market in southeastern Australia. After early exploratory fishing, commercial harvesting commenced in 1915, when the New South Wales government imported three steam trawlers from England. These early vessels

---

[47]This section is based on R.D.J. Tilzey (ed.), *The South East Fishery: A Scientific Review with Reference to Quota Management* (Canberra, 1994).

used otter-trawl gear, a similar method to that used by modern trawlers operating in the region today. The government constructed additional vessels until the fleet was privatised in 1923. Expansion continued until about 1929, when seventeen steam trawlers were operating, but numbers declined slowly to fourteen vessels just prior to the Second World War. Danish-seine vessels began fishing in 1933 and took about twenty percent of the catch before the onset of hostilities. The navy requisitioned fishing vessels during the war, leading to a substantial decrease in catch and fishing effort in the region. By 1943, only one steam trawler and a few Danish-seine vessels were operating. Immediately after the war effort expanded, and between 1946 and 1954 ten to twelve steam trawlers operated in the fishery. This fishing mode declined from 1954 to 1961, when the last steam trawler ceased operations. The fleet consisted of Danish- seine vessels alone from 1961 to 1971, when the first modern diesel otter trawler commenced in the fishery. Since that time, the number of Danish-seine vessels has declined slowly to thirty, while the fleet of diesel trawlers has grown, with 105 active in 1991. In the latter year, an ITQ system was introduced. This has led to the consolidation of quota holdings by multi-vessel operators and to a slight decrease in the number of vessels fishing.

The fishery was largely concentrated in shelf waters at depths between fifty and 200 metres. The targets were tiger flathead (*Neoplatycephalus richardsoni*), jackass morwong (*Nemadactylus macropterus*) and redfish (*Centroberyx affinis*). In the 1970s, the spawning run of gemfish (*Rexea solandri*) was discovered in slope waters between 300 and 400 metres, adding another important species to the fishery. The total annual landed catch fluctuated between 1000 and 7500 tonnes per annum between 1918 and 1961, increasing to a peak of 62,269 tonnes in 1990. This was mainly due to the discovery of orange roughy (*Hoplostethus atlanticus*) on the shelf slope at depths of more than 500 metres.

Figure 4 provides a temporal overview of the history of the southeast trawl fisheries. The shelf and slope ecosystems were essentially untouched by humans prior to the twentieth century. Since then, exploitation by trawl fishing has been substantial. The area of the New South Wales trawl grounds is equal to about two-thirds of the Irish Sea, but about four times the Irish Sea catch was taken there in 1927 by far fewer boats and men.[48] A wealth of data exists on this fishery, but since little is available in computerized form, analyses of the historical record have been limited. But the evidence indicates that the fishery has had substantial impacts on both fish biomass and community structure. Catch and effort data from a group of steam trawlers have been retrieved for three distinct

---

[48]W.S. Fairbridge, "The Effect of the War on the East Australian Trawl Fishery," Journal of the Council for Scientific and Industrial Research, XXI, No. 2 (1948), 75-98.

periods.[49] The catch on the shelf for these steam trawlers declined by a factor of almost four between the early 1920s and the late 1930s and has remained low (figure 5). There was an increase in catch rates in the early 1940s as a consequence of the decrease in fishing effort during the Second World War. Associated with the decline in catch rates was the changing mix of species taken. This is illustrated in figure 6, in which the species composition of the catch from these same steam trawlers is plotted for the principal fishing ground of Botany Bay, near Sydney. It shows both the appearances of new species and the disappearance, or near disappearance, of species that were originally caught.

The trawl fishery expanded into the deeper waters of the continental slope during the 1970s with the introduction of the modern diesel otter trawlers. The general history of the fishery on the slope was similar to that on the shelf. Research surveys carried out by the New South Wales Fishery Department indicate a decline in biomass of the species taken in the trawl. Accompanying this decline were major shifts in species composition.[50]

There is a wealth of information on both the shelf and slope fisheries spanning the entire period of exploitation (table 2). Thus, there are detailed catch and effort data that can provide evidence on catch weight, composition and location on a haul-by-haul basis from the start of exploitation in the early 1900s. Yet a "modern," systematic statistical data collection system for this fishery was only established in 1985. Accordingly, much of the data on this system are historic in nature and require appropriate historical and analytical interpretation. For example, an estimate of total catches by major commercial species from this fishery does not exist. Thus, to fully understand these systems, and the ecological consequences of exploitation, there is a need for historical research on what is seemingly a relatively recent fishery.

In terms of marine systems with a history of intense exploitation, it is perhaps unique that scientific research survey data exist for both the shelf and slope ecosystems prior to the period of commercial exploitation. In addition, scientific survey data have been collected at various times subsequently. These survey data can provide important snapshots which greatly enhance the prospects of understanding the ecological changes that have occurred.

---

[49]N.L. Klaer and R.D.J. Tilzey, "Catalogue and Analysis of South East Fishery Historic Data," Final Report to the Australian Fisheries Research and Development Corporation, Project No. 90/023, 1996.

[50]K.J. Graham, B.R. Wood and N.L. Andrew, *The 1995-97 Survey of Upper Slope Trawling Grounds between Sydney and Gabo Island (and Comparisons with the 1976-77 Survey)* (Sydney, 1997).

Figure 3
The Location of the Southeast Australian Trawl Fishery
Prior to the 1930s

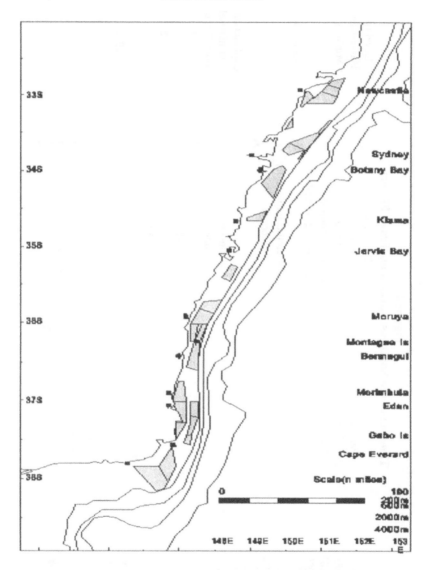

Source: A.N. Colefax, "A Preliminary Investigation of the Natural History of the Tiger Flathead (*Neoplatycephalus macrodon*) on the South-Eastern Australian Coast," *Procedings of the Linnaean Society of New South Wales*, LIX (1934), 71-91.

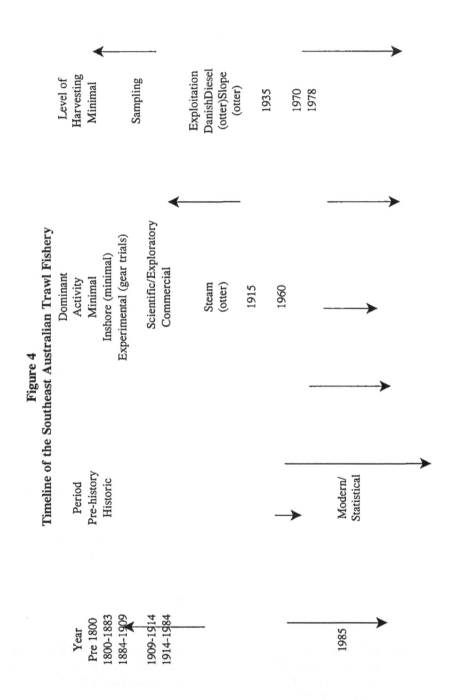

Figure 4
Timeline of the Southeast Australian Trawl Fishery

Table 2
Data Sources on the Southeast Australian Trawl Fishery

Endeavour Surveys (not computerized)
        1909-1914 log books exist (not computerized)
Historic Commercial Logs collected by CSIRO (computerized)
        1918-1923 records for 12,041 hauls
        1937-1943 records for 31,266 hauls
        1951-1957 records for 31,266 hauls
Red Funnel Company Records (computerized)
        1952-1961 Three vessel logbooks (7464 hauls)
        1938-1959 Daily by vessel Landing Records (5572)
        1946-1957 Daily Radio Reports (8016)
Sydney and Melbourne Fish Market (not computerized prior to 1970, partially computerized since then)
        1915-present
Historic Non-computerized Personal Commercial Logbook Data
Recollections, diaries, interview etc. from participants
Numerous Research Surveys
Environmental Data
Official Logbook collected and maintained by AFMA (computerized)
NSW State Fisheries Observer
        1991-present

Figure 5
Time Trend of Catch Rates (Total Catch Divided by Number of Hauls) from Steam Trawler Haul-by-Haul Data, 1918-1957

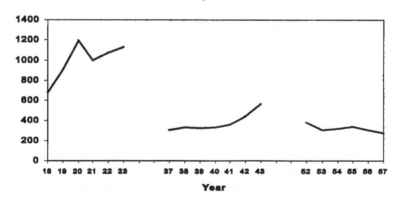

CPUE (kg/haul)

Source: Values taken from N.L. Klaer and R.D.J. Tilzey, "Catalogue and Analysis of South East Fishery Historic Data," Final Report to the Australian Fisheries Research and Development Corporation, Project No. 90/023, 1996, with corrected values for 1918-1923.

Figure 6
Species Composition Changes on Shelf Fishing Grounds off Sydney,
1918-1957

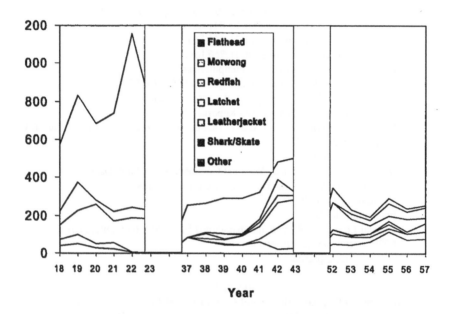

CPUE (kg/haul)

Note: Catch per unit of effort (CPUE) per species from the haul-by-haul steam trawl
records for the area off Botany Bay.

Source: See text.

While there is a wealth of potentially valuable information on these
systems, a substantial portion of the historic data has not been assembled, let
alone computerised, so that comprehensive analyses have yet to be undertaken.
In addition, since these fisheries are relatively recent, there is the opportunity to
recover important information from individuals who can recall their early years.
While a welter of information relates to the entire history of exploitation,
preliminary analyses indicate that the fishery has had a major impact on the
biological community. These factors suggest that this system can provide
valuable insights into the ecological processes of ecosystem structure and the
impact of human exploitation.

New Zealand Statistics

Following New Zealand's establishment as a Crown Colony in 1840, the Colonial Secretariat prepared *Blue Books* for the Colonial Office between 1841 and 1852.[51] Although New Zealand was granted responsible government in 1856, the first yearbook was not published until 1875, while New Zealand's first *Official Yearbook* did not appear until 1893. The post of government statistician, attached to the Registrar-General's Office, was created in 1910. A separate Census and Statistics Office was founded in 1915, initially as a branch of the Department of Internal Affairs before it was transferred in 1931 to the Department of Industries and Commerce. A separate Census and Statistics Department was created in 1936. It was renamed the Department of Statistics in 1956 and in 1994 renamed Statistics New Zealand.[52]

Some statistics on whaling were collected as early as 1829, but compared to Australia's official publications, they provided only limited data on fisheries.[53] For example, the *New Zealand Official Yearbook* included the following summary data on the fishing industry in 1911:[54]

Total employment	1442
Fish exports	NZ $55,216
Oysters exported	NZ $4,356
Whalebone and whale oil exported	NZ $13,848
Total exports	NZ $73,420
Imports of fish and fish products	NZ $179,318

From 1937 onwards, the individual landings from each licensed fishing boat were collected to enable the Chief Inspector of Fisheries to present a more detailed report.[55] For example, in 1937 the quantities and values of fish caught in Auckland, the main fishing port, were as follows:

---

[51]"Sources of Statistical Information," *New Zealand Official Yearbook* (Wellington, 1953), 1016-1026.

[52]McLintock (ed.), *Encyclopaedia of New Zealand*, III, 310-311.

[53]"Sources of Statistical Information," 1016.

[54]*New Zealand Official Yearbook* (Wellington, 1912), 854. All financial data prior to 1967 has been converted to dollars on the basis that £1 = NZ $2.

[55]*New Zealand Official Yearbook* (Wellington, 1938), 464-465.

Scale fish	159,371 cwt	NZ$235,900
Crayfish	1325 cwt	NZ$3108
Mussels	7019 sacks	NZ$3654

Scale fish include snapper, tarakihi, trevally, flounder, gurnard, hapuku, dory, kingfish, blue cod, barracouta, mullet, and garfish, but weights for individual species were not shown.

Before the Second World War, New Zealand fisheries statistics suffered from many deficiencies, including inaccurate species identification, misreporting of catches and inaccurate industry returns. For example, common names of fish were often applied differently in each region. The kingfish of Dunedin was the hake of Wellington and the lemon fish of Auckland.[56] The failure to apply correct scientific names led to confusion in data collection and fish marketing.

Currently, the New Zealand Ministry of Fisheries collects detailed, high-quality data on fifty species for the purpose of assessment of stocks and sustainability.[57] The data on each of these species are subdivided into commercial fisheries, recreational fisheries, Maori customary fisheries, illegal catch and other sources of mortality, although in most cases data are available only on the first two categories. Additional information is provided on biology, stocks and areas, stock assessment and status of the stocks. This is a rich data source, although the amount of historical information varies between the different fish species. In the case of grouper, for example, extensive data are available. The first recorded landings measured about 1500 tonnes in 1936. Catches fell during the Second World War due to reduced effort, but grew rapidly after 1945 to a peak of 2000 tonnes in 1949. They then declined slowly to about 1300 tonnes in the mid-1970s, and increased quickly to a record of 2700 tonnes in 1983-1984. A total allowable catch (TAC) was introduced in 1986-1987 but has rarely been exceeded in any fishery; since 1992-1993 total landings have averaged between sixty and seventy percent of TAC. Since the early 1990s, catches have stabilized at around 1450 tonnes. By contrast, the Japanese and Russians developed the fishery for hoki only in the early 1970s and, while it quickly grew to become the largest fishery in New Zealand – in 1996-1997 the reported catch was a massive 269,000 tonnes – catch statistics are not available before 1969.

---

[56]E.P. Neale, *Guide to New Zealand Official Statistics* (Auckland, 1938), 50.

[57]See New Zealand, Ministry of Fisheries, "Summary of the Assessments of the Sustainability of Current Total Allowable Commercial Catch (TACCs) and Recent Catch Levels and Status of the Stocks for the 1999-2000 Fishing Year" (http://www.fish.govt.nz/sustainability/index.html).

Conclusion

This paper provides a preliminary review of sources for the historical study of fisheries. A comprehensive survey requires further library and archival research, especially in New Zealand. With the exception of whaling and pearling, the secondary literature on the history of fishing is still relatively limited. This lacuna is especially apparent in the case of New Zealand, where there is a pressing need for further research on the history of fishing. While the HMAP project is primarily concerned with the collection of statistics for the reconstruction of fish populations and testing ecological hypotheses, it also has the potential to help correct such gaps in the historical literature.

British colonial authorities regularly collected statistical information on the progress of colonies, and as a result both Australia and New Zealand have rich sources of statistical information dating from the early years of settlement. But while sealing and whaling briefly attained economic importance, fishing was a relatively minor activity and consequently received little coverage in official statistics. This paper shows, however, that there is sufficient statistical material to establish long time series for the total catch, imports and exports and some major species. The case study of the Southeast Australian trawl fishery shows that there are detailed catch and effort data from the start of fishing in the early 1900s and therefore good prospects of reconstructing the complete history of exploitation. But the available statistics are primarily by weight rather than value, of variable quality and subject to many qualifications. Nevertheless, together with other forms of evidence, such as manuscripts, oral evidence, and anthropological and scientific research, they provide the basis for an improved understanding of the development of fisheries resources in Australia and New Zealand. Moreover, incorporating the Australasian data into the HMAP project can go some way to meeting Pontecorvo's ambitious plea for a global fishery database.

Examining Cetacean Ecology
Using Historical Fishery Data

Tim D. Smith

Abstract

Historical fisheries data have been used to examine aspects of the ecology of many cetacean species. Examples are discussed for three populations of whales: eastern North Pacific gray whales, North Atlantic long-finned pilot whales and North Atlantic humpback whales. Reconstructions of historical gray whale populations have improved understanding of population size relative to pre-exploitation sizes but have also suggested the possibility of long-term ecological change. Changes in the size composition of pilot whales indicate long-term fluctuations in production of young whales, implying greater vulnerability to ecological and environmental changes than is often assumed for top predators. Changes in the spatial distribution of humpback whales on breeding grounds over time have suggested differential recovery of components of populations or, alternatively, major shifts in the location of breeding grounds. There is potential to resolve some of these issues through the systematic analysis of historical sources, notably whaling logbooks and governmental records, through the examination of historical environmental sources, and through the interpretation of evidence from archeological investigations.

Introduction

Whalers have recorded various aspects of whaling operations and catches in ships' logbooks. Such information has been used with commercial and governmental records of the numbers of whaling voyages and fishery production to determine levels of catch and fishing effort. In some shore-based fisheries, written records were kept of individual whales in order to regulate the distribution of catches. Further, aboriginal catches of several species of whales have been inferred from archival and archeological records. Systematic information can often be constructed from such historical material even though the reliability and consistency of the sources vary greatly. This technique has been used to estimate historical abundances, to explore spatial distribution patterns and to examine changes in population productivity.

207

Several attempts have been made to use data extracted from such sources to reconstruct historical population sizes.[1] Two approaches have been used, both motivated by a need to determine the effects of harvesting. One approach back-calculates historical population size from current estimates of population size, taking account of total catches and population productivity. This requires reliable estimates of historical catches and current abundance, and knowledge of the species' biology. Methods for measuring current population size based on sighting surveys and on sight-resight studies have been developed, and these have provided reasonably precise estimates in some cases.[2] Estimating rates of productivity is more difficult, but has been done by using a series of abundance estimates and by using catalogues of individually-marked animals.[3] Estimates of catches, abundance and productivity have been used as a base for gauging historic population size over spans of decades for dolphins, and of more than a century for bowhead whales.[4] The second method of estimating historic population size is to compare abundance

---

[1]In 1983, the International Whaling Commission conducted a general workshop on the use of historical whaling records. See M. Tillman and G. Donovan (eds.), "Historical Whaling Records," *Reports of the International Whaling Commission*, Special Issue 5 (1983).

[2]For sighting surveys, see S. Buckland, *et al.*, *Distance Sampling: Estimating Abundance of Biological Populations* (London, 1993); for sight-resight studies, see G. Seber, *The Estimation of Animal Abundance and Related Parameters* (New York, 1982). Ratios of standard errors to estimates of abundances of ten percent have been obtained for minke whales using sighting surveys by T. Schweder, *et al.*, "Abundance of Northeastern Atlantic Minke Whales, Estimates for 1989 and 1995," *Reports of the International Whaling Commission*, XLVII (1996), 453-483. A ratio of seven percent for humpback whales for sight-resight studies has been obtained by T.D. Smith, *et al.*, "An Ocean-Basin Wide Mark-Recapture Study of the North Atlantic Humpback Whale (*Megaptera novaeangliae*)," *Marine Mammal Science*, XV (1999), 1-32.

[3]For bowhead whales, see J.E. Zeh, *et al.*, "Rate of Increase, 1978-1988, of Bowhead Whales, *Balaena mysticetus*, Estimated from Ice-Based Census Data," *Marine Mammal Science*, VII (1991), 105-122; for humpback whales, see J. Barlow and P. Clapham, "A New Birth-Interval Approach to Estimating Demographic Parameters of Humpback Whales," *Ecology*, LXXVIII (1997), 535-546.

[4]For dolphins, see T. Smith, "Changes in Size of Three Dolphin (*Stenella spp.*) Populations in the Eastern Tropical Pacific," *Fishery Bulletin*, LXXXI (1983), 1-13; and P. Wade, "Abundance and Population Dynamics of Two Eastern Pacific Dolphins, *Stenella attenuata* and *Stenella longirostris*" (Unpublished PhD thesis, University of California, San Diego, 1994). For bowhead whales, see A. Raftery, G. Givens and J. Zeh, "Inference from a Deterministic Population Dynamics Model for Bowhead Whales (with discussion)," *Journal of the American Statistical Association*, XC (1995), 402-430.

indices with total catches. Both types of data are often available, with abundance indices expressed in terms of sightings or catches per unit of fishing or searching effort.[5] This approach is possible when historical sources contain information on the fishing process as well as catches, and it can be combined with the back-calculation approach as well.

The seasonal and spatial distribution of whales is often revealed in whaling records. Examination of where harvesting occurred, and where it did not occur even though the opportunity would have presented itself had whales been present, has suggested major changes over centuries in the habitat of some species. The size composition of catches can often be used to infer changes in reproductive rates, especially in connection with information on harvesting intensity and environmental records. Analyses of historical whaling data have also raised important ecological questions. This paper focuses on three such analyses, relating respectively to the eastern North Pacific gray whale, the North Atlantic long-finned pilot whale and the North Atlantic humpback whale. The state of the analysis is discussed in each case, with attention also afforded to the ways that such data might be further used to test ecological hypotheses.

Eastern North Pacific Gray Whale: Back-Calculating Abundance from Historical Records of Total Catch

The gray whale population in the eastern North Pacific migrates between feeding grounds in the Bering Sea and breeding lagoons in the Gulf of California. The population was hunted to extremely low levels during the late nineteenth and early twentieth centuries when commercial catches greatly exceeded the long-standing aboriginal catches. Commercial catches were estimated by using whaling logbook data, while aboriginal catches were gauged using historical reviews and interviews with "scholars in the field." In recent decades, population size has been calculated directly from shore-based migration-corridor sighting surveys.[6] Survey results suggest that the population has increased from roughly 10,000 whales in the 1960s to more than 20,000 in the 1990s.

To determine the present status of the gray whale population relative to historical levels, various population models have been fitted to the estimated

---

[5]For example, fishing days on Pacific sperm whaling grounds have been extracted from logbooks and adjustments have even been made for the effects of weather on efficiency. See M. Tillman and J. Breiwick, "Estimates of Abundance for the Western North Pacific Sperm Whale based upon Historical Whaling Records," *Reports of the International Whaling Commission,* Special Issue 5 (1983), 257-269.

[6]S. Buckland, "Estimated Trends in Abundance of California Gray Whales from Shore Counts, 1967/68 to 1987/88," International Whaling Commission, Scientific Committee Meeting Document SC/A90/G9 (1990).

catches and abundances using knowledge of the species' life history. But the estimated trajectories have been inconsistent with recent population abundance increases. This inconsistency may be due either to inadequacy in the estimated historical catches or to changes in ecological carrying capacity.[7] The first of these possibilities could be tested by improving the quality of the historical catch estimates by further evaluating logbooks and other materials that reflect commercial and aboriginal catches. This could also help overcome a tendency to err on the side of underestimation.[8] More systematic evaluation of aboriginal catches using anthropological and historical tools would also be fruitful.[9] The second possibility could be tested by evaluating other data relating to the state of the ecosystem. For instance, there is scope for more detailed examinations of long term changes in benthic fauna prey and in environmental conditions, while evidence of changes in whale vital rates, such as the possible decline in pregnancy rates as suggested by recent aboriginal catches, might prove instructive.

Faroe Islands Pilot Whale Drive Fishery: Shore-Based Catch Distribution Records

The long-finned pilot whale is widely distributed in the North Atlantic and has been subject to coastal "drive fisheries" in several areas.[10] The drive fishery in the

---

[7]See Anon., "Report on the Special Meeting of the Scientific Committee on the Assessment of Gray Whales," *Reports of the International Whaling Commission*, XLIII (1993), 241-259, and references therein for previous modelling attempts and a summary of alternative hypotheses that are consistent with the results.

[8]Mitchell identified several uncertainties in the historical aboriginal and commercial catch estimates, including changes in rates of oil production, direct and indirect calf mortality, and "killed but lost." E. Mitchell, "Comments on Gray Whale Catch Statistics," *Reports of the International Whaling Commission*, XLIII (1993), 256-257. N. Friday and T.D. Smith, "The Effect of Age and Sex Selective Harvest Patterns on Whale Populations," International Whaling Commission, Scientific Committee Meeting Document, SC/52 (2000), identified the potentially greater biological significance of catches focussed on mothers and calves.

[9]See Tillman and Donovan, "Historical Whaling Records," for a more general discussion of this point. E. Mitchell and R. Reeves, "Aboriginal Whaling for Gray Whales in the Eastern North Pacific Ocean," International Whaling Commission, in press, present the results of a further review of aboriginal catches, which suggests levels roughly fifty percent greater than previously estimated.

[10]A.G. Abend and T.D. Smith, "Review of Distribution of the Long-Finned Pilot Whale (*Globicephala melas*) in the North Atlantic and Mediterranean," *NOAA Technical Memorandum* NMFS-NE-117 (1999).

Faroe Islands, for example, has continued for several centuries. Faroese catches are allocated among those people helping with each drive or *grind*, and the allocation is based upon records made by local authorities at the time of the catch. These records include both numbers of whales and a volume-related measure of individual size, termed *skinn*. Records exist back to the 1600s and appear to be complete since the 1860s. These data have been analysed by using deterministic population models which, with the long term persistence of this fishery, suggest that current catch levels may be sustainable.[11]

Fluctuations in the number and size of the animals harvested indicate long-term aperiodic changes. Further analyses of the size composition reported in the historical records hint that the proportion of young animals caught has varied substantially over the decades.[12] These changes imply that reproductive success has varied substantially. Such variability, however, is inconsistent with the assumptions used in the deterministic population models considered in evaluating the effects of the fishery and hence draws into question the conclusions noted above. From an ecological perspective, the catch data suggest that pilot whales may be more susceptible to environmental or ecological variability than previously suspected of cetaceans.[13] Additional analyses of the historical environmental data (e.g., the North Atlantic Oscillation Index) would assist efforts to understand more fully the role of ecological variability in cetacean life history.

North Atlantic Humpback Whale: Commercial Logbooks as Evidence of Distribution Patterns

The population of humpback whales in the North Atlantic was greater than 10,000 animals in the early 1990s, and in the aggregate is increasing. The population has a complex life history that involves both breeding grounds in the western and

---

[11]For a description of the fishery and the major scientific study in the late 1980s, see the papers in G. Donovan, C. Lockyer and A. Martin (eds.), "Biology of the Northern Hemisphere Pilot Whales," *Reports of the International Whaling Commission,* Special Issue 14 (1993). See also Anon., "Report."

[12]D. Bloch and L. Lastein, "Modelling the School Structure of Pilot Whales in the Faroe Islands, 1832-1994," in A. Blix, L. Walløe and O. Ulltang (eds.), *Whales, Seals, Fish and Man* (Amsterdam, 1995), 499-508; and T.D. Smith, D. Bloch and S. Brault., "Long Term Fluctuations in Age and Sex Composition of Pilot Whales (*Globicephala melas*) in the Faroe Islands," forthcoming.

[13]Hjort's early perspective that "the renewal of the stock is bound up with the fate of a limited progeny whom nature has safeguarded in various ways against the many causes of mortality" has been frequently cited in subsequent studies. J. Hjort, "Whales and Whaling," *Hvalradets Skrifter,* VII (1933), 7-29.

eastern Caribbean islands and in the Cape Verde Islands and feeding grounds spread across the northern North Atlantic.[14] Animals using the separate feeding grounds appear to mix, at least on the western Caribbean breeding grounds where dense aggregations have been observed, but have very high fidelity to their mothers' feeding grounds.[15] Although humpback whales have been observed in recent years in both the Cape Verde and eastern Caribbean islands, comparable aggregations to those in the western Caribbean have not been reported.[16] The rate of increase of the population components in the feeding grounds varies, with the rate in the Gulf of Maine exceeding five percent annually, while that in West Greenland appears to be near zero.[17]

Humpbacks were subjected to intensive exploitation in the eighteenth century, and commercial logbook data have been extracted to determine population history.[18] Analysis of these data suggests that the total population size prior to documented exploitation could have been less than 5000 animals. Such a low value is inconsistent with present abundance estimates of twice that level unless there have been major changes in the carrying capacity of the ecosystem. This possible inconsistency could be due at least in part to incomplete historical catch data. Further analysis of historical logbooks may help determine the extent of this problem, especially accessing historical records in various Caribbean

---

[14]See Smith, *et al.*, "Ocean-basin Wide Mark-Recapture," for abundance. A preliminary analysis of photographic identification data over the 1970s and 1980s suggests a positive rate of increase (personal communication, Peter Stevick, University of St. Andrews).

[15]See references in Smith, *et al.*, "Ocean-basin Wide Mark-Recapture;" and in P. Palsbøll, *et al.*, "Genetic Tagging of Humpback Whales," *Nature*, CCCLXXXVIII (1997), 767-769.

[16]F. Reiner, M.E. Dos Santos and F.W. Wenzel, "Cetaceans of the Cape Verde Archipelago," *Marine Mammal Science*, XII (1996), 434-443; and H.E. Winn, R.K. Edel and A.G. Taruski, "Population Estimate of the Humpback Whale (*Megaptera novaeangliae*) in the West Indies by Visual and Acoustic Techniques," *Journal of the Fisheries Research Board of Canada*, XXXII (1975), 499-506.

[17]For the Gulf of Maine, see Barlow and Clapham, "New Birth-Interval Approach;" for West Greenland, see F. Larsen and P. Hammond, "Distribution and Abundance of West Greenland Humpback Whales," International Whaling Commission, Scientific Committee Working Paper SC/52 (2000).

[18]Mitchell and Reeves, "Catch History;" and W.S. Price, "Whaling in the Caribbean: Historical Perspective and Update," *Reports of the International Whaling Commission*, XXXV (1985), 413-420.

nations. But other changes in the environment that might affect carrying capacity could also be at least partially responsible for this apparent inconsistency.

Data from whaling logbooks can also be used to examine other ecological phenomena. For example, the seasonal and geographical evidence in logbooks suggests that most historical winter whaling was in the eastern Caribbean and the Cape Verde Islands.[19] In contrast, no evidence of whaling in the western Caribbean has been found, even though some whaling ships provisioned near the largest current breeding ground in the western Caribbean.[20] Given the large historical catches from the eastern Caribbean and the recent recovery of the North Atlantic population in total, the apparent low abundance of humpback whales in the eastern Caribbean and the Cape Verde Islands is surprising. Indeed, a recent spatially-expansive survey conducted in the eastern Caribbean by using both passive acoustic and visual sighting methods detected the presence of humpbacks throughout the survey area, although no aggregations comparable to those in the western Caribbean were detected.[21]

These observations suggest that either whales using the western and eastern Caribbean and Cape Verdean breeding grounds have recovered differentially from previous depletion, or there has been a shift in breeding grounds since the time of peak exploitation.[22] Such shifts are possible in migratory large mammals and have been suggested for humpback whales in the Pacific.[23] Further

---

[19]R.R. Reeves, *et al.*, "Historical Occurrence of Humpback Whales in the Eastern and Southern Caribbean Sea, based on Data from American Whaling Logbooks," International Whaling Commission, Scientific Committee Working Paper SC/52 (2000); and R.R. Reeves, P.J. Clapham and S. Wetmore, "Humpback Whaling at the Cape Verde Islands, and a Summary of Humpback Catches elsewhere in the Eastern North Atlantic," International Whaling Commission, Scientific Committee Working Paper SC/52 (2000).

[20]D.K. Mattila, *et al.*, "Occurrence, Population Composition and Habitat Use of Humpback Whales in Samana Bay, Dominican Republic," *Canadian Journal of Zoology*, LXXII (1994), 1898-1907.

[21]S. Swartz, "Visual and Acoustic Survey of Humpback Whales (*Megaptera novaeangliae*) in the Eastern Caribbean – The Windwards: Preliminary Findings," International Whaling Commission, Scientific Committee Working Paper SC/52/IA1 (2000).

[22]P.J. Clapham and L.T. Hatch, "Defining Management Units for Animal Populations: The Example of Whales," International Whaling Commission, Scientific Committee Meeting Document, SC/52/ID2 (2000).

[23]For example, colonization and re-colonization of breeding islands have been documented for several pinniped species (personal communication, Richard Merrick, Northeast Fisheries Science Center, Woods Hole, MA). On humpbacks, see L. Herman,

historical data will enable researchers to test the differential recovery and breeding ground shift hypotheses more rigorously.

Conclusions

The examples given raise questions about possible changes in carrying capacity, breeding habitat and breeding success. Such changes are contrary to the single-species, density-dependent paradigm that is usually adopted when considering such higher trophic level species as cetaceans and raise questions about how to think about what is "natural." For example, a reconstruction of Caribbean sea turtles suggested the ecological effects of populations of grazing mega-fauna (e.g., turtles and manatees) on Caribbean coral reefs may have been similar to grazing herbivores on the Serengeti plains.[24] Studies of historical fisheries data as suggested here are important if we are to avoid the "shifting baseline syndrome,"[25] which takes the form that "everyone, scientists included, believed that the way things were when they first saw them is natural."[26]

It will be necessary to make better use of such data to overcome this temporal limitation in perspective. This will require more intensive historical and archeological research, as well as the collaboration of whale biologists, ecologists, climatologists and historians.

"History of the Hawaiian Humpback Whale," in Report On a Workshop On Problems Related to Humpback Whales (*Megaptera Novaeangliae*) in Hawaii, US Marine Mammal Commission, Pb-280 794, MMC-77/03 (1978), 25-28.

[24]J. Jackson, "Reefs since Columbus," *Coral Reefs*, XVI, supplement 1 (1997), 23-32.

[25]D. Pauly, "Anecdotes and the Shifting Baseline Syndrome of Fisheries," *Trends in Ecology and Evolution*, X (1995), 430.

[26]Jackson, "Reefs since Columbus."

Epilogue

Poul Holm, David J. Starkey and Tim D. Smith

The workshop at which the papers that comprise this volume were presented also generated a research agenda for the "History of Marine Animal Populations (HMAP)" project. This agenda, in turn, formed the basis of a proposal that subsequently attracted financial support from the Alfred P. Sloan Foundation (New York City). Having commenced in January 2001, the HMAP initiative provides an historical dimension to the "Census of Marine Life," a decade-long research program designed to assess and explain the diversity, distribution and abundance of marine life in the world's oceans.

HMAP is an international, interdisciplinary research project. It is based in three institutions: the University of Southern Denmark, the University of Hull, UK, and the University of New Hampshire, USA. These HMAP Centres are responsible for co-ordinating and managing the research efforts of more than fifty historians, ecologists and ecosystem modelers working in thirty-one institutions in eighteen countries. The researchers have been recruited to assemble and analyse historical and paleo-ecological data relating to human harvesting in seven distinct spatial contexts over the past 500 years. Their case studies, which will be completed in 2003, will not only yield information on the impact of historic and contemporary fishing activity on ecosystem dynamics but will also develop methodologies that can be applied more widely to other elements of the marine environment. The seven case studies are: Southwest Pacific (southeast Australian shelf and slope fisheries, New Zealand shelf fisheries); Southwest African Shelf (clupeid fisheries in a continental boundary current system); California Current (clupeid fisheries in a continental boundary current system); Northwest Atlantic (Gulf of Maine, Newfoundland-Grand Banks, Greenland cod fisheries); Norwegian, North and Baltic Seas (multinational cod, herring and plaice fisheries); White and Barents Seas (Russian and Norwegian herring, salmon and cod fisheries); and Whaling Worldwide (historical and twentieth-century whaling in all oceans).

The HMAP workshop identified a series of ecological and historical hypotheses that were incorporated into the design of the project. Using historical and paleo-ecological data, the researchers, in conjunction with the HMAP Centres, will test the following six hypotheses: that historical records can be used to infer fish population and community structure, after accounting for anthropogenic factors; that anthropogenic changes in fishery patterns include changes in socio-economic-political-demographic factors, technology, numbers of vessels and individuals in the fishery, and changes in knowledge; that environmental forcing causes changes in abundance and/or spatial distribution; that fishing mortality has significant impacts on population abundance and/or spatial distribution; that changes in energy flows across trophic structure are due to environmental change or fishing mortality; and that diversity of marine animals has declined due to exploitation and habitat loss.

Work on the research programme is currently in its early stages. But it is anticipated that the findings of the HMAP case studies will make a major contribution to knowledge and understanding of the complex, delicate and important relationship between human societies and the marine environment.

Printed and bound by CPI Group (UK) Ltd, Croydon, CR0 4YY

27/10/2024

14580406-0002